I0484299

Agency Leads
Krisa Arzayus
National Oceanic and Atmospheric Administration
Marilyn Buford
U.S. Department of Agriculture
James Butler
National Oceanic and Atmospheric Administration
Roger Dahlman
Department of Energy
David Hofmann
National Oceanic and Atmospheric Administration
Fred Lipschultz
National Science Foundation
Peter Murdoch
U.S. Geological Survey
Ed Sheffner
National Aeronautics and Space Administration
Diane Wickland
National Aeronautics and Space Administration

For More Information
U.S. Climate Change Science Program
1717 Pennsylvania Avenue, NW, Suite 250
Washington, DC 20006 USA
+1.202.223.6262 (voice)
+1.202.223.3065 (fax)
http://www.climatescience.gov/

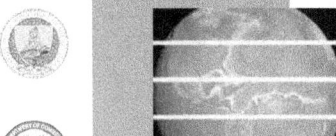

1. DESCRIPTION OF TOPIC, AUDIENCE, INTENDED USE, AND QUESTIONS TO BE ADDRESSED

1.1. Introduction

The carbon cycle chapter of the *Strategic Plan for the U.S. Climate Change Science Program* (CCSP) describes a plan to produce "…a series of increasingly comprehensive and informative reports about the status and trends of carbon emissions and sequestration," each to be called a State of the Carbon Cycle Report (SOCCR). The Carbon Cycle Interagency Working Group's (CCIWG) Terms of Reference (TOR)[1] for a first SOCCR elaborated this in June 2003, saying that what is envisioned is "…a series of reports on the state of the carbon cycle designed to provide accurate, unbiased, and policy-relevant scientific information concerning the carbon cycle to a broad range of stakeholders. The two broad objectives for a State of the Carbon Cycle Report are (1) to summarize scientific knowledge about carbon cycle properties and changes, and (2) to provide scientific information for decision support and policy formulation concerning carbon." The first SOCCR will be CCSP Synthesis and Assessment Product (SAP) 2.2.

The carbon cycle chapter of the *CCSP Strategic Plan* describes a long-term vision to regularly produce a comprehensive report on the state of the global carbon cycle within 10 years and projects that a near-term, prototype report focused on North America can be produced within 2 years. SAP 2.2 will summarize substantive information about North America's carbon budget. It also will serve as a prototype for future enhancement and extension to a global report. Subsequent reports are expected to evolve based on the lessons learned in producing earlier reports.

1.2. Topic and Content

SAP 2.2 will provide a synthesis and integration of the current knowledge of the North American (including land, atmosphere, and adjacent oceans) carbon budget and its context within the global carbon cycle. In a format useful to decisionmakers, it will (1) summarize our knowledge of carbon cycle properties and changes relevant to the contributions of and impacts[2] upon the United States and the rest of the world, and (2) provide scientific information for U.S. decision support focused on key issues for carbon management and policy.

---

[1] The *Terms of Reference for a First State of the Carbon Cycle Report* can be found at <http://www.carboncyclescience.gov>. It was prepared by the CCIWG in consultation with its Science Steering Group, completed in May 2003, and posted on the web in June 2003.

[2] The term "impacts" as used in this prospectus refers to specific effects of changes in the carbon cycle, such as acidification of the ocean, the effect of increased CO_2 on plant growth and survival, and changes in concentrations of carbon in the atmosphere. It is not used as a shortened version of "climate impacts," as was adopted for the *Strategic Plan for the U.S. Climate Change Science Program*.

SAP 2.2 will address carbon emissions; natural reservoirs and sequestration; rates of transfer; the consequences of changes in carbon cycling on land and the ocean; effects of purposeful carbon management; effects of agriculture, forestry, and natural resource management on the carbon cycle; and the socio-economic drivers and consequences of changes in the carbon cycle. It will cover North America's land, atmosphere, inland waters, and adjacent oceans. It will include an analysis of North America's carbon budget that will document the state of knowledge and quantify best estimates (i.e., consensus, accepted, official) and uncertainties. This analysis will provide a baseline against which future results from the North American Carbon Program (NACP) can be compared. SAP 2.2 will be coordinated with other CCSP synthesis and assessment products as appropriate, especially 2.1 (*Scenarios of Greenhouse Gas Emissions and Atmospheric Concentrations and Review of Integrated Scenario Development and Application*) and 3.1 (*Climate Models: An Assessment of Strengths and Limitations for User Applications*). More specifically, SAP 2.2 will:

- Quantify current information on sources and sinks and associated uncertainties related to the buildup of carbon dioxide and methane in the atmosphere. For example, it will provide best estimates of the contribution of carbon dioxide emissions from combustion of fossil fuels in North America to changes in global atmospheric CO_2 concentrations for recent decades. Discussion of future changes in fossil fuel emissions will be limited to existing scenarios because scenarios are the central element of the work being done under SAP 2.1.
- Discuss and assess current accepted projections of the future of the North American carbon budget, including uncertainties in projected fossil fuel emissions and the impact of policy and technology scenarios on those emissions.
- Provide current estimates, with the associated uncertainties, of the fractions of global and North American fossil-fuel carbon emissions being taken up by North America's ecosystems and adjacent oceans.
- Provide current, best available answers to specific questions about the North American carbon budget relevant to carbon management policy options. The

questions will be identified through early and continuing dialogue with SAP 2.2 stakeholders. The answers will include explicit characterization of uncertainties.
- Identify where NACP-supported research will reduce current uncertainties in the North American carbon budget and where future enhancements of NACP research can best be applied to further reduce critical uncertainties.
- Describe and characterize the carbon cycle as an integrated interactive system, using innovative graphics to depict the carbon cycle in ways that are easily understandable.

1.3. Audience

The audience for SAP 2.2 includes scientists, decisionmakers in the public sector (Federal, State, and local governments), the private sector (carbon-related industry, including energy, transportation, agriculture, and forestry sectors; and climate policy and carbon management interest groups), the international community, and the general public. This broad audience is indicative of the diversity of stakeholder groups interested in knowledge of carbon cycling in North America and of how such knowledge might be used to influence or make decisions. Not all scientific information needs of this broad audience can be met in this first synthesis and assessment product, but the scientific information to be provided will be designed to be understandable by all. The primary users of SAP 2.2 are likely to be officials involved in formulating climate policy, individuals responsible for managing carbon in the environment, and scientists involved in assessing and/or advancing the frontier of knowledge.

1.4. Intended Use

SAP 2.2 will be used (1) as a state-of-the-art assessment of our knowledge of carbon cycle properties and changes relevant to the contributions of and carbon-specific impacts upon the United States in the context of the rest of the world; (2) as a contribution to relevant national and international assessments; (3) to provide the scientific basis

for decision support that will guide management and policy decisions that affect carbon fluxes, emissions, and sequestration; (4) as a means of informing policymakers and the public concerning the general state of our knowledge of the global carbon cycle with respect to the contributions of and impacts on the United States; and (5) as a statement of the carbon cycle science information needs of important stakeholder groups. For example, well-quantified regional- and continental-scale carbon source and sink estimates, error terms, and associated uncertainties will be available for use in U.S. climate policy formulation and by resource managers interested in quantifying carbon emissions reductions or carbon uptake and storage. It is expected that participating scientists will publish parallel research articles in peer-reviewed science journals. These research articles will augment SAP 2.2 as a baseline against which to compare future NACP results and as input to future Intergovernmental Panel on Climate Change (IPCC) assessments. Senior managers and the general public will use the Executive Summary of SAP 2.2 and the SOCCR web site—created to support SAP 2.2 development—to improve their overall understanding of the U.S. role in Earth's carbon budget and to gain perspective on what is and is not known.

1.5. Questions to be Addressed

Questions to be addressed by SAP 2.2 follow:
- What is the carbon cycle and why should we care?
- How do North American carbon sources and sinks relate to the global carbon cycle?
- What are the primary carbon sources and sinks in North America, and how are they changing and why?
- What are the direct, non-climatic effects of increasing atmospheric CO_2 or other changes in the carbon cycle on the land and oceans of North America?
- What are the options and measures implemented in North America that could significantly affect the North American and global carbon cycles (e.g., North American sinks and global atmospheric CO_2 concentrations)?
- How can we improve the application of scientific information to decision support for carbon management and climate decisionmaking?

These questions are starting points for producing SAP 2.2; they were developed by the proposed SAP 2.2 Coordinating Team (see Section 3) and refined at the first stakeholders workshop. The draft outline of major sections of the report (see Attachment 1) elaborates on how they will be addressed in the report.

2. CONTACT INFORMATION: E-MAIL AND TELEPHONE FOR RESPONSIBLE INDIVIDUALS AT THE LEAD AND SUPPORTING AGENCIES

The lead agencies for SAP 2.2 are the Department of Energy (DOE), the National Oceanographic and Atmospheric Administration (NOAA), and the National Aeronautics and Space Administration (NASA); the responsible individuals are Dr. Roger Dahlman, Dr. David Hofmann and Dr. James Butler, and Dr. Diane Wickland and Mr. Ed Sheffner, respectively. For legal purposes only, including those of the Information Quality Act (IQA) and Federal Advisory Committee Act (FACA), NOAA has been designated the single lead agency for SAP 2.2 and, as such, is responsible for ensuring compliance with NOAA's Information Quality Guidelines (<http://www.noaanews.noaa.gov/stories/iq.htm>) and the Office of Management and Budget's *Information Quality Bulletin for Peer Review* (<http://www.whitehouse.gov/omb/inforeg/peer2004/peer_bulletin.pdf>). Dr. Krisa Arzayus of NOAA is the point of contact for matters concerning IQA and FACA compliance. Supporting agencies are the U.S. Department of Agriculture (USDA), U.S. Geological Survey (USGS), and National Science Foundation (NSF); the responsible individuals are Dr. Marilyn Buford, Mr. Peter Murdoch, and Dr. Fred Lipschultz, respectively.

CCSP Agency	Agency Leads
NOAA	Krisa Arzayus
	Krisa.Arzayus@noaa.gov
	(301) 713-2465
USDA	Marilyn Buford
	mbuford@fs.fed.us
	(703) 605-5176

NOAA	James Butler	
	James.H.Butler@noaa.gov	
	(303) 497-6898	
DOE	Roger Dahlman	
	Roger.Dahlman@science.doe.gov	
	(301) 903-4951	
NOAA	David Hofmann	
	David.J.Hofmann@noaa.gov	
	(303) 497-6966	
NSF	Fred Lipschultz	
	flipschu@nsf.gov	
	(703) 292-7701	
USGS	Peter Murdoch	
	pmurdoch@usgs.gov	
	(518) 285-5663	
NASA	Ed Sheffner	
	Edwin.J.Sheffner@nasa.gov	
	(202) 358-0239	
NASA	Diane Wickland	
	Diane.E.Wickland@nasa.gov	
	(202) 358-0245	

This group of lead and supporting agency representatives has been designated the "Agency Executive Committee" (AEC) and will be hereafter referred to as such. The CCIWG has formally approved that the AEC will fulfill the role of the "Executive Committee" envisioned in the SOCCR TOR. As NOAA leads SAP 2.2 through the expert peer review, public comment, CCSP, and National Science and Technology Council (NSTC) approvals, and dissemination processes, it will consult regularly with the AEC, and prior to initiating each major step in the overall process.

3. LEAD AUTHORS: REQUIRED EXPERTISE OF LEAD AUTHORS AND BIOGRAPHICAL INFORMATION FOR PROPOSED LEAD AUTHORS

In 2004, the CCIWG received, conducted a peer review, selected, and funded a proposal from a team of scientific experts to prepare the first SOCCR (SAP 2.2). The proposal was unsolicited and was received after the CCIWG's TOR for SOCCR was made publicly available. NASA, NOAA,

DOE, and NSF agreed to provide the funding. The selected proposal was from the Battelle Memorial Institute (operating as UT-Battelle, LLC out of the Oak Ridge National Laboratory), an outside contractor. The funding award has been set up such that the U.S. Government will not exert management or control over the activities of the contractor nor will U.S. Government officials play a role in selecting authors, holding meetings, setting the agenda, or drafting the final report. NOAA has determined that this approach to produce SAP 2.2 does not require a FACA committee. The lead authors and their roles are:

Dr. Anthony King
  Overall Lead
  Oak Ridge National Laboratory
Dr. Lisa Dilling
  Stakeholder Interaction Lead
  University of Colorado/
  National Center for Atmospheric Research
Dr. David Fairman
  Stakeholder Interaction
  Consensus Building Institute, Inc.
Dr. Richard A. Houghton
  Scientific Content (Land Use)
  The Woods Hole Research Center
Dr. Gregg Marland
  Scientific Content (Emissions)
  Oak Ridge National Laboratory
Dr. Adam Rose
  Scientific Content (Economics)
  The Pennsylvania State University
Dr. Thomas Wilbanks
  Scientific Content (Human Dimensions)
  Oak Ridge National Laboratory

Their activities will be coordinated by:

Mr. Gregory Zimmerman
  Project Coordinator
  Oak Ridge National Laboratory

These individuals will be responsible for organizing and outlining SAP 2.2 and for its final content and submission to NOAA. They will identify chapter authors, coordinate all the inputs to SAP 2.2, and lead the overall synthesis and integration of the report. They will provide oversight and editorial review of individual chapters and will, with the chapter authors, prepare any overview chapters and the Executive Summary. In order to minimize confusion with the group of chapter authors, this group of lead authors and the Project Coordinator will hereafter be referred to as the "SAP 2.2 Coordinating Team." Their biographies are provided in Attachment 2.

The responsibility for writing each individual chapter of SAP 2.2 will be assigned to one or more scientific experts in the topic area of the chapter; this person (or persons) will be designated the lead chapter author(s). The chapter authors will be recognized leaders in their fields, drawn from the wide and diverse scientific community of North America and the world, as well as other qualified stakeholder groups. Qualifications that will be recognized are the quality and relevance of current publications in the peer-reviewed literature pertaining to their chapter topics, past or present positions of leadership in the topic fields, and other documented experience and knowledge of high relevance. Lead chapter authors will be responsible for the review and synthesis of current knowledge and production of text. They will be responsible for recruiting well-qualified contributing authors in their areas of expertise and responsibility. Chapter authors will be responsible for ensuring that scientific expert, stakeholder, and public review comments on their chapters are reflected in the final report. All authors will be listed in association with their contributions (e.g., chapters) in the final report.

The following lead chapter authors have been contacted and have agreed to participate in drafting the SOCCR (SAP 2.2):

Dr. Richard Birdsey, U.S. Forest Service
Dr. Scott Bridgham, University of Oregon
Dr. Ken Caldeira, Lawrence Livermore National Laboratory
Dr. Francisco Chavez, Monterey Bay Aquarium
  Research Institute
Dr. Rich Conant, Colorado State University
Dr. Kenneth Davis, The Pennsylvania State University

Dr. Lisa Dilling, University of Colorado/
  National Center for Atmospheric Research
Dr. David Fairman, Consensus Building Institute, Inc.
Dr. Chris Field, Carnegie Institution
Dr. David Greene, Oak Ridge National Laboratory
Dr. Erik Haites, Margaree Consultants
Dr. Burke Hales, Oregon State University
Dr. Richard A. Houghton, Woods Hole Research Center
Dr. Elizabeth Huber-Sannwald, Institute for Scientific
  and Technological Research of San Luis Potosi
Dr. Mark Jaccard, Simon Frazer University
Dr. Jennifer Jenkins, University of Vermont
Dr. Mark Johnston, Saskatchewan Research Council
Dr. Gregg Marland, Oak Ridge National Laboratory
Dr. James McMahon, Lawrence Berkeley National Laboratory
Dr. Ron Mitchell, University of Oregon
Dr. Stephen Pacala, Princeton University
Dr. Diane Pataki, University of California-Irvine
Dr. Keith Paustian, Colorado State University
Dr. Patricia Romero Lankao, Metropolitan Autonomous
  University – Xochimilco
Dr. Adam Rose, The Pennsylvania State University
Dr. Jorge Sarmiento, Princeton University
Dr. Taro Takahashi, Lamont-Doherty Earth Observatory,
  Columbia University
Dr. Pieter Tans, National Oceanic and Atmospheric
  Administration
Dr. Charles Tarnocai, Agriculture and Agri-Food Canada
Dr. Thomas Wilbanks, Oak Ridge National Laboratory
Dr. Steven Wofsy, Harvard University

Their biographies are provided in Attachments 2 and 3. The SAP 2.2 Coordinating Team has discussed the draft chapter outline and candidate chapter authors in its initial consultations with science, government, private sector, and other stakeholders, and provided opportunities for comments and additional nominations during these consultations and from the public through the CCSP and SOCCR (SAP 2.2) web posting and comment processes.

The chapter author's assignment to lead a specific topical chapter has been determined as part of this process. Lead and contributing chapter author selections were made to

ensure a balance of scientific and technical expertise and that disparate views that have significant scientific support are represented. Final authorship decisions were made by the SAP 2.2 Coordinating Team, communicated to NOAA and the AEC, and posted on the SOCCR (SAP 2.2) web site. The lead authors for each chapter are identified in Attachment 1.

4. STAKEHOLDER INTERACTIONS

A process for engaging important stakeholder groups and establishing an ongoing dialogue with them will be a priority activity. Stakeholder involvement is essential to ensure *transparency* (open access to information on SAP 2.2), *feedback on relevance* (review and comment on the SAP 2.2 process and verification that information produced by SAP 2.2 will be useful), and *credibility* (recognition by the stakeholders of the scientific validity and independence of SAP 2.2). These activities will be the responsibility of the SAP 2.2 Coordinating Team. Their plan includes "a structured dialogue between scientists and stakeholders to identify and clarify information needs of managers and decisionmakers" as the first of two major SAP 2.2 tasks.

The process of engaging stakeholders requires first establishing a meaningful, two-way dialogue. The SAP 2.2 Coordinating Team notes in its proposal that "the initial design and context are critically important and that the framing process requires great care." The SAP 2.2 Coordinating Team's plan for a structured dialogue with stakeholders involves a partnership with the Consensus Building Institute, Inc.—an organization that has broad experience working with diverse stakeholder communities in the energy and environmental sectors. A multistage process has been planned to provide access and information exchange (see Section 9 for the proposed timeline).

Significant activities have already been conducted to seek stakeholder input and to scope the report. They were conducted as SOCCR activities, without reference to

SAP 2.2. These activities were used to prepare this prospectus and its attachments, including:

- An initial draft outline of the SOCCR was produced by the SOCCR Coordinating Team and delivered to the AEC on 30 September 2004.
- A stakeholder assessment involving in-depth interviews and discussions with approximately 30 representatives of key stakeholder communities (e.g., scientists, policymakers, policy advocates, and carbon-related industries) was initiated 1 October 2004. Representatives of key stakeholder constituencies were identified by taking advantage of existing stakeholder contacts, processes such as CCSP's web posting and public comment process, inputs from individuals providing information for the update to the Voluntary Greenhouse Gas Registry, CCIWG member's knowledge of key policymakers and groups, and referrals from the stakeholders contacted. Inputs were assessed in order to narrow focus to stakeholders needs in a few key areas, then to conduct in-depth interviews with stakeholders in those areas. This assessment resulted in a November 2004 *State of the Carbon Cycle Report Stakeholder Assessment Report.*
- A web site for SOCCR (<http://www.ucar.edu/soccr>) was developed and put online in October 2004, with information on progress and planning for the SOCCR. A listserve mailing list was established to distribute electronic information about SOCCR and contains over 300 accounts.
- A First Stakeholders Workshop for the SOCCR was held at the Key Bridge Marriott hotel in Arlington, Virginia, 15-16 November 2004. Twenty-seven participants from industry, academia, environmental interest organizations, scientists/researchers, and decisionmakers from the Federal government attended the workshop. A primary objective of this First Stakeholders Workshop was to seek input on how well the 30 September 2004 draft outline addressed scientific, policy, business, and other interests and concerns. The workshop resulted in the creation of a revised outline responsive to the interests/ needs of the stakeholders. The workshop also identified additional opportunities for future stakeholder involvement throughout the development of the SOCCR.

- The draft outline produced at the First Stakeholder Workshop (Attachment 1) was posted on the SOCCR web site on 19 November 2004 for a public comment period of 30 days ending 19 December 2004. Notice of the availability of the SOCCR outline for comment was e-mailed to all interviewees, workshop participants, candidate chapter authors, and individuals on the SOCCR listserve shortly after posting on the web. A number of comments were received through the automated web site and considered according to the *Guidelines for Producing Synthesis and Assessment Products*. The comments received and the lead authors' responses to them have been posted on the SOCCR web site.

- A "sounding board" composed of individuals of widely recognized expertise and stature in carbon cycle research has been established to provide input to the SOCCR Coordinating Team primarily on scientific/ technical issues in preparing the report.

- A Town Hall meeting on the SOCCR (entitled *The State of the Carbon Cycle Report (SOCCR): Integrating Scientific Synthesis and Assessment with Stakeholders Interests and Issues*) was held 16 December 2004, as part of the 2004 American Geophysical Union (AGU) Fall Meeting in San Francisco, California.

- A "Joint Authors-Stakeholders Workshop" bringing the lead chapter authors of the draft SOCCR together with a diversity of stakeholders was held at the Crystal City Marriott hotel in Arlington, Virginia, on 24 October 2005. Lead chapter authors and stakeholders from public and private research institutions and governmental organizations attended the workshop. A primary objective of the Joint Workshop was to seek input from stakeholders on the relevance of the SOCCR chapter material for their decisionmaking processes at an early stage in the document's formulation. Based upon feedback from the stakeholders, the Workshop resulted in a modification of the structural content of the SOCCR chapters to make the report more consistent across the entire document. The stakeholders also identified a few instances where discussions of additional topics could easily be added to improve the document. The day following this workshop (25 October 2005) the lead chapter authors met to reflect on and respond to the results of the previous

day's dialogue with stakeholders, and to discuss their individual scientific perspectives in relation to the integration of their respective chapters into the overall SOCCR report.

- On 8 December 2005, the Coordinating Team hosted a Town Hall (entitled *The State of the Carbon Cycle Report (SOCCR): Integrating Scientific Synthesis and Assessment with Stakeholder Interests and Issues*) as part of the Fall 2005 AGU meeting. The Coordinating Team reported on the progress on the SOCCR, described the timeline for its scheduled completion, and answered questions on content and process.

One additional stakeholders workshop will be conducted to foster communication, establish interactions among stakeholders and SAP 2.2 authors, and develop inputs to shape the content of SAP 2.2. Throughout the development of SAP 2.2, inputs from stakeholders will be communicated to the SAP 2.2 chapter authors so that the report can be revised and refined. The SAP 2.2 Coordinating Team has taken advantage of CCSP's posting and review process to both identify stakeholders and capture additional inputs from them. Stakeholder inputs that cannot be incorporated into SAP 2.2 will be captured and summarized so they can be used to inform future *State of the Carbon Cycle Reports*.

5. DRAFTING PROCESS (INCLUDING MATERIALS TO BE USED IN PREPARING THE PRODUCT)

The SAP 2.2 Coordinating Team has discussed the draft chapter outline in its initial consultations with science, government, private sector, and other stakeholders, and provided opportunities for comments and additional nominations during these consultations and from the public through the CCSP and SOCCR (SAP 2.2) web posting and comment processes. The SAP 2.2 Coordinating Team is responsible for the detailed outline of SAP 2.2 and making final decisions about the scope and full content of the report. The SAP 2.2 Coordinating Team is responsible for ensuring the report is well integrated, balanced, and responsive. The SAP 2.2 Coordinating Team plans to achieve the scientific synthesis through compilation and

analysis of the relevant scientific literature and available databases. Since SAP 2.2 will be completed during the initial stages of NACP, much of the information for SAP 2.2 will, by necessity, be derived from publications of many independent investigations and may consider portions of North America or may subset North America from larger geographical analyses. Many decisions will be required about how to handle disparate information. These issues were discussed with chapter authors at both the first (May 2005) and second (October 2005) authors' workshops.

Many data sets required for SAP 2.2 are already available at data archives such as the NOAA Climate Monitoring Diagnostics Laboratory (CMDL), the DOE Carbon Dioxide Information Analysis Center (CDIAC), and the NASA Distributed Active Archive Centers (DAACs). However, some of the scientific questions raised by SAP 2.2 will require further data compilation, synthesis, and integration efforts. The SAP 2.2 Coordinating Team will compile a central tabulation of referenced and supporting data, including links to available data, documentation, and contact information for data that are not easily accessible. The use of unpublished data will be discouraged for SAP 2.2. If any such data should be proposed for use, approval will be sought consistent with the *Guidelines for Producing CCSP Synthesis and Assessment Products*. SAP 2.2 will also require tabulation of data that are not purely numerical. As described above, the effective coordination of SAP 2.2 will depend on a systematic and regularly updated tabulation of the activities of ongoing related programs, with contact information and links to relevant web sites. The proper documentation of in-text citations will require compilation of a substantial web-accessible bibliographic database.

All authors will be provided with NOAA's Information Quality Guidelines as specified in the *Guidelines for Producing CCSP Synthesis and Assessment Products*, which will include compliance with the overall Office of Management and Budget guidelines, *OMB Guidelines for Ensuring and Maximizing the Quality, Objectivity, Utility, and Integrity of Information Disseminated by Federal Agencies and the Information Quality Bulletin for Peer Review*.

The authors of SAP 2.2 will be expected to emphasize accuracy and precision of numerical information, confidence levels, characterization of uncertainties, and transparency of original data and model sources. SAP 2.2 will provide a clear discussion of uncertainties and how uncertainties may be reduced, preferably through a section of each chapter in which measurements, model results, or combinations of data and models occur. Numerical values will be accompanied by measures of uncertainty (e.g., ± x units or percent). Where the uncertainty cannot be quantified, an explanation or justification will be given. Statements that are vague will be avoided. All data used in SAP 2.2 (or linked by a SAP 2.2-related website) will be clearly documented, including data source and other information needed to evaluate information.

To ensure consistency and thoroughness in the treatment of uncertainties across all chapters of SAP 2.2, the SAP 2.2 Coordinating Team will maintain regular oversight of overall data and information quality as presented in workshops and in draft text.

6. REVIEW

NOAA will ensure that SAP 2.2 is reviewed at all stages as specified in the *Guidelines for Producing CCSP Synthesis and Assessment Products* and consistent with the Information Quality Act and *Information Quality Bulletin for Peer Review*, that comments and other feedback are provided to the SAP 2.2 Coordinating Team for response, and that comments and responses are documented and made publicly available.

6.1. During Drafting Period

The SAP 2.2 Coordinating team plans to post on the SOCCR (SAP 2.2) web site the list of authors and all drafts of the outline, chapters, and complete report, with a mechanism for providing comments through the web site. The SAP 2.2 Coordinating Team will also establish a process and standards for ongoing information quality review.

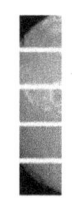

6.2. Expert Review of First Draft

NOAA will coordinate a formal, external expert peer review of the first draft, drawing from the national and international communities of scientific and technical experts and following the highest standards of rigor in peer review. Expert peer reviewers will be deemed qualified through their record of scholarly publication in the topic areas of SAP 2.2 and/or comparable experience and accomplishment that are well-documented. NOAA will draw from all CCIWG agencies' lists of qualified expert peer reviewers and will solicit suggestions for expert peer reviewers from the scientific community and other stakeholders through the CCSP and SAP 2.2 web posting and comment processes. The public is invited to nominate expert reviewers to participate in the peer review of the draft SAP 2.2. Nominations should be sent to Dr. Krisa Arzayus at <Krisa.Arzayus@noaa.gov> by 30 April 2006, and must include the potential reviewer's contact information, curriculum vitae, and list of publications. Reviewer selections will be made to ensure a balance of scientific and technical expertise and that disparate views that have significant scientific support are considered appropriately. The reviewers will include experts with knowledge of the types of information and level of technical detail that will make the report useful to decisionmakers and other stakeholders. Peer reviewers who are Federal employees will be subject to Federal requirements governing conflict of interest [see 18 U.S.C. 208, 5 C.F.R. Part 2635 (2004)]. Reviewers who are not Federal employees will be screened pursuant to the National Academy of Sciences policy for committee selection with respect to conflict of interest. The charge to reviewers, their names and affiliations, and unattributed solicited comments will be posted on NOAA's Information Quality Act web site (<www.osec.doc.gov/cio/oipr/info_qual.html>) and linked and/or replicated on the CCSP web site (<www.climatescience.gov>).

The expert review of the first draft will be conducted by a minimum of 15 expert peer reviewers who will submit comments similar to those solicited as part of a journal peer review. Mail-in evaluations will be requested from these reviewers. Reviewers will be asked to use the following questions in formulating their comments:

- Is the scope and intent for the synthesis and assessment product clearly described in the report? Are all aspects of this charge fully addressed? Do the authors go beyond this charge or their expertise?
- Are the conclusions and recommendations adequately supported by evidence, analysis, and argument?
- Are uncertainties or incompleteness in the evidence explicitly recognized?
- Are the data and analyses handled competently? Are statistical methods applied appropriately?
- Are the report's exposition and organization effective? Is the title appropriate?
- Is the report fair and appropriately balanced?
- Is the report's tone impartial and devoid of special pleading?
- Are any of the report's findings based on value judgments or the collective opinions of the authors? If so, is this acknowledged, and are scientifically defensible reasons given for reaching those judgments?
- Does the executive summary concisely and accurately describe the key findings and recommendations? Is it consistent with the other sections of the report?
- What other significant improvements, if any, might be made in the report?

NOAA does not plan to convene a peer review panel, but reserves the right to do so (by either calling a meeting or holding a teleconference) if conflicting comments or detailed technical considerations need to be resolved prior to providing feedback to the SAP 2.2 Coordinating Team. The reviews, as submitted, will be made available to the SAP 2.2 Coordinating Team. The Coordinating Team and the lead authors of SAP 2.2 will revise the draft report by incorporating comments and suggestions from the reviewers as they deem appropriate based on their scientific judgment. The authors will acknowledge significant contributions made by expert reviewers, as applicable. The Coordinating Team will prepare a written response to the peer reviewers' comments explaining its agreement or disagreement with the views of the peer reviewers; the actions taken in response to the peer review; and the reasons why those

actions respond to the peer reviewers' key concerns. This response will be made publicly available. The expert peer review will be conducted during a two-month period to start in May 2006.

6.3. Public Review of the Second Draft

After revision, the second draft SAP 2.2 will be released for public comment. The public comment period will be 45 days. Following this comment period, the authors will prepare a third draft of the report, taking into consideration the comments submitted during the public comment period. The scientific judgment of the authors will determine responses to the technical comments. All comments submitted during the public review will be made publicly available. The public comment period will begin in September 2006.

6.4. CCSP and NSTC Review of the Third Draft

Once the revisions to the second draft are complete, the SAP Coordinating Team will submit the third draft of the synthesis and assessment product to NOAA. Once NOAA determines that the report conforms to CCSP and IQA guidelines, it will submit the draft product and a compilation of the comments received to the CCSP Interagency Committee. If the CCSP Interagency Committee determines that further revision is necessary, their comments will be sent to NOAA to seek consideration and resolution by the Coordinating Team and lead authors. If needed, the National Research Council (NRC) will be asked to provide additional scientific analysis to bound scientific uncertainty associated with specific issues.

If the CCSP Interagency Committee review determines that no further revisions are needed and that the report has been prepared in conformance with the *Guidelines for Producing CCSP Synthesis and Assessment Products* and the Information Quality Act (including ensuring objectivity, utility, and integrity as defined in 67 FR 8452), they will submit the report to the NSTC for clearance. Clearance will

require the concurrence of all members of the Committee on Environment and Natural Resources. The CCSP Interagency Committee will be responsible for seeing that comments generated during the NSTC review are addressed. They will consult with NOAA and the authors to develop an appropriate response. If the synthesis and assessment product should need to be revised, the revisions will be written by the SAP 2.2 Coordinating Team and/or chapter authors and then routed back through NOAA and the CCSP Interagency Committee to the NSTC. All comments generated by the NSTC review and the responses to them will be made publicly available.

After clearance and prior to publication, the AEC, CCIWG, Coordinating Team, and all authors will be given the opportunity to examine the final report. If at this stage, or any earlier stage in this process, an individual author cannot accept the outcomes of the writing, review, and revision process, they will be accorded the opportunity to withdraw their name from the publication.

7. RELATED ACTIVITIES, INCLUDING OTHER NATIONAL AND INTERNATIONAL ASSESSMENT PROCESSES

As a near-term report, SAP 2.2 will utilize, to the maximum extent possible, the information available from existing data, programs, and related activities in the United States and internationally. SAP 2.2 will be coordinated with related work in a way that does not duplicate previous and ongoing assessments. Coordination with the NACP will be necessary to ensure that the most current information is available to scientists and stakeholders contributing to SAP 2.2 and so that NACP benefits from the scientific baseline and assessment of stakeholder needs for scientific information that SAP 2.2 will establish. SAP 2.2 will be both informed by and used as an input to relevant national and international assessments.

A particular concern is the development of partnerships with international groups whose interests overlap those of SAP 2.2. Although SAP 2.2 will be a U.S. report, the information in SAP 2.2 must reflect international scientific understanding. It is imperative that SAP 2.2 be coordinated with ongoing international efforts to avoid duplication of effort, to maximize effectiveness, and to ensure that the most up-to-date integrated science is presented in a global context. The SAP 2.2 Coordinating Team will ensure that relevant international scientific bodies are informed of the intent and progress of the SAP 2.2 and will seek to harmonize its efforts with ongoing relevant work of such bodies.

The SAP 2.2 Coordinating Team will establish informal communications with participants in IPCC, the Global Carbon Project (GCP), and national programs in Canada and Mexico. The schedule for the next IPCC assessment report is such that the results of SAP 2.2 will not be available in time to be incorporated. However, informal communications among the authors of the two activities will ensure that knowledge of the most up-to-date and reliable information and analyses is exchanged.

8. COMMUNICATIONS: PROPOSED METHOD OF PUBLICATION AND DISSEMINATION OF THE PRODUCT

Once NSTC clearance has been obtained, NOAA will coordinate publication and release of SAP 2.2. Financial support for the production and distribution of the final SAP 2.2 will come from the Federal government agencies participating in the CCIWG. SAP 2.2 will be printed and hardcopies will be made available through the CCSP Office; it will also be made available electronically on both the CCSP and SOCCR (SAP 2.2) web sites. The published report will follow the standard format for all CCSP synthesis and assessment products.

An interactive, high-quality web site has been developed for SOCCR (SAP 2.2) and will be used to make SAP 2.2 and a wide variety of information about it available to all stakeholders and the general public. The web site will serve multiple functions: complementing the printed version of SAP 2.2, allowing worldwide access to SAP 2.2 from any

internet location; expanding SAP 2.2 content in a fashion that will be especially useful to the research community by allowing users to click on links for further information, references, notes, etc., under specific sections of the text; linking to U.S. agency and international carbon cycle science and management websites (providing a web portal to highlight all of the existing, ongoing work); and providing an interactive way for users to comment on their experience of SAP 2.2 and how it might be made more useful in the future.

Opportunities for offering information to the SAP 2.2 Coordinating Team will be broadly disseminated in scientific and other public venues. The SAP 2.2 Coordinating Team, chapter authors, and other participants in SAP 2.2 will publicize the SAP 2.2 process widely. The purposes are to disseminate information about the process and to persuade key stakeholders to participate and use the SAP 2.2 report as an aid to management and decisionmaking. A package of material will be created for all those involved in SAP 2.2 to use as they travel in their ongoing professional work. The SOCCR (SAP 2.2) web site will be publicized at scientific meetings, to agency representatives, and at other appropriate venues (e.g., carbon sequestration meetings). The web site will explain the process of the SAP 2.2, and list information as it is approved for release. There will be an opportunity for comments to be logged on that site, and records will be kept of all comments as well as the responses to those comments.

9. PROPOSED TIMELINE

Activity	Months From Start	Estimated Completion Date
Start work	0	1 September 2004
Submit draft outline to AEC	1	1 October 2004
Identify and initiate consultations with stakeholders	1.5	16 October 2004

First Stakeholders Meeting	2.5	15-16 November 2004
Establish SOCCR web site	2.5	15 November 2004
CCSP posts prospectus for public review	5	2 February 2005
Public review period for prospectus ends	6	7 March 2005
First Chapter Authors Workshop	8.5	16-17 May 2005
NOAA guidance on FACA as applied to SAP 2.2 process	13.5	20 October 2005
Joint Authors-Stakeholders Workshop	13.5	24-25 October 2005
Chapter authors' materials/ manuscripts compiled	17.5	13 February 2006
CCSP posts revised, final prospectus	17.5	14 February 2006
Expert reviewer nominations due	20	30 April 2006
Submit draft SAP 2.2 to NOAA	20.5	12 May 2006
Complete expert peer review of draft SAP 2.2	22	7 July 2006
Deliver revised SAP 2.2 to NOAA	24	1 September 2006
Post revised SAP 2.2 for public review and comment	24	8 September 2006
Third Stakeholders Meeting	25	September or October 2006
Public review and comment period closes	26	31 October 2006
Complete and deliver SAP 2.2 to NOAA	29	31 January 2007
CCSP and NSTC review completed and SAP 2.2 released	30	March 2007

ATTACHMENT 1. CHAPTER STRUCTURE OF SYNTHESIS AND ASSESSMENT PRODUCT 2.2: THE STATE OF THE CARBON CYCLE REPORT

EXECUTIVE SUMMARY (authors: SOCCR Coordinating Team)

Chapter 1. Introduction to the Report's Purpose, Scope, and Structure: What is the carbon cycle and why should we care? (authors: SOCCR Coordinating Team)
(In Brief: The report is designed to provide accurate, unbiased, and policy-relevant scientific information concerning the carbon cycle to a broad range of stakeholders, including scientists and non-scientists. Stakeholders for the SAP 2.2 have expressed an interest in both synthetic information as well as detailed information for particular types of ecosystems or activities. Accordingly, Part I is an interdisciplinary, integrated synthesis aimed at answering overarching questions on the nature and status of the North American carbon cycle. Part I also establishes the global context, including atmosphere and oceans, for the continental-scale North American carbon budget.   Parts II and III include chapters with a more sectoral or disciplinary focus. Part II addresses energy, industry and waste management activities in North America. Part III addresses the land and water ecosystems of the continent. Chapters in Part II and III are intended to reach both scientist and non-scientist stakeholders who wish to review information on a specific sector in greater detail.   Workshops and author communication across chapters will ensure that information is not redundant and also remains consistent across the sectoral chapters of Parts II and III and the cross-cutting, synthetic chapters of Part I.)

PART I: THE CARBON CYCLE IN NORTH AMERICA

Chapter 2. How do North American carbon sources and sinks relate to the global carbon cycle? (lead authors: Chris Field (Coordinating Lead), Burke Hales, Jorge Sarmiento, and others)

Chapter 3. What are the primary carbon sources and sinks in North America, how are they changing and why? (lead authors: Steve Pacala and Steve Wofsy (Coordinating Leads), Ken Davis, Burke Hales, Richard Houghton, Pieter Tans, and others)

Chapter 4. What are the options and measures that could significantly affect the carbon cycle? (lead authors: Erik Haites (Coordinating Lead), Ken Caldeira, Patricia Romero Lankao, Adam Rose, Tom Wilbanks)

Chapter 5. How can we improve the application of scientific information to decision support for carbon management and climate decision-making? (lead authors: Lisa Dilling and Ron Mitchell (Coordinating Leads), David Fairman)

PART II: ENERGY, INDUSTRY, AND WASTE MANAGEMENT ACTIVITIES

Overview of Part II: *Title* (to be determined); (author: Gregg Marland)

Chapter 6: Energy Extraction and Conversion (lead author: Gregg Marland)

Chapter 7: Transportation (lead author: David Greene)

Chapter 8: Industry and Waste Management (lead author: Mark Jaccard)

Chapter 9: Buildings (lead author, James McMahon)

PART III: LAND AND WATER SYSTEMS

Overview of Part III: *Title* **(to be determined); (author: Richard Houghton)**

Chapter 10. Agriculture, Grassland, Shrubland and Arid Lands (lead authors: Keith Paustian, Rich Conant)

Chapter 11. Forests (lead authors: Mark Johnston, Jennifer Jenkins, Richard Birdsey, and Elisabeth Huber-Sannwald)

    Introduction and Summary

    A. Boreal Forests
    B. Temperate Forests
    C. Tropical Forests

Chapter 12. Carbon Cycle in Permafrost Regions (i.e., Boreal, Subarctic and Arctic Areas) of North America (lead author: Charles Tarnocai)

Chapter 13. Non-Permafrost Wetlands (lead author: Scott Bridgham)

Chapter 14. Human Settlements and the North American Carbon Cycle (lead author: Diane Pataki)

Chapter 15. Aquatic Carbon, Coastal Management, and Ocean Basins (lead authors: Francisco Chavez and Taro Takahashi)

ATTACHMENT 2. BIOGRAPHIES OF SOCCR COORDINATING TEAM
(I.E., SAP 2.2 "LEAD AUTHORS")

Anthony W. King

Environmental Sciences Division
Oak Ridge National Laboratory
P.O. Box 2008
Oak Ridge, TN 37831-6335
Tel: (865) 576-3436; Fax: (865) 574-2232

Education

1978	B.S.	Zoology, Arkansas State University
1981	M.S.	Biology, Arkansas State University
1986	Ph.D.	Ecology, University of Tennessee, Knoxville

Research interests

Terrestrial ecosystems as part of the global Earth system, ecosystem and land-surface processes at landscape, regional, and global scales, climate-ecosystem feedbacks, carbon and water cycle modeling, land-use change, spatially structured population dynamics and modeling, theory of scale and system organization in ecology, model sensitivity and uncertainty analysis, model evaluation.

Employment History

| 1992-present | Research Staff Member, Environmental Sciences Division, Oak Ridge National Laboratory |
| 1987-1992 | Research Associate, Environmental Sciences Division, Oak Ridge National Laboratory |

Selected Publications

Amthor, J.S., J. M. Chen, J.S. Clein, S.E. Frolking, M.L. Goulden, R.F. Grant, J.S.Kimball, A.W. King, A.D. McGuire, N.T. Nikolov, C.S. Potter, S. Wang and S.C. Wofsy. 2001. Boreal forest CO_2 and evapotranspiration predicted by nine ecosystem process models: inter-model comparisons and relationships to field measurements. Journal of Geophysical Research 106:33,623-33,648.

Potter, C.S., S. Wang, N.T. Nikolov, A.D. McGuire, J. Liu, A.W. King, J.S. Kimball, R.F. Grant, S.E. Frolking, J. Clein, J.M.Chen and J.S. Amthor. 2001. Comparison of boreal ecosystem model sensitivity to variability in climate and forest site parameters. Journal of Geophysical Research 106:33,671-33,688.

King, A.W., W.M. Post and S.D. Wullschleger. 1997. The potential response of terrestrial carbon storage to changes in climate and atmospheric CO_2. Climatic Change 35:199-227.

King, A.W., W.R. Emanuel, S.D. Wullschleger and W.M. Post. 1995. In search of the missing carbon sink: a model of terrestrial biospheric response to land-use change and atmospheric CO_2. Tellus 47B:501-519.

King, A.W., R.V. O'Neill and D.L. DeAngelis. 1989. Using ecosystem models to predict regional CO_2 exchange between the atmosphere and the terrestrial biosphere. Global Biogeochemical Cycles 3:337-361.

Jager, H.I., T.L. Ashwood, B.L. Jackson and A.W. King. 2000. Spatial uncertainty analysis of ecological models. Proceedings of the 4th International Conference on Integrating GIS and Environmental Modeling (GIS/EM4): Problems, Prospects, and Research Needs. Banff, Alberta, Canada, September 2-8, 2000.

Jager, H.I., W.W. Hargrove, C.C. Brandt, A.W. King, R.J. Olson, J.M.O. Scurlock and K.A. Rose. 2000. Constructive contrasts between modeled and measured climate responses over a regional scale. Ecosystems 3:396-411.

Post, W.M., A. King and S.D. Wullschleger. 1997. Historical variations in terrestrial biospheric carbon storage. Global Biogeochemical Cycles 11:99-109.

King, A.W., W.R. Emanuel and W.M. Post. 1992. Projecting future concentrations of atmospheric CO_2 with global carbon cycle models: simulating historical changes in atmospheric CO_2. Environmental Management 16:91-108.

Post, W.M., T.-H. Peng, W.R. Emanuel, A.W. King, V.H. Dale and D.L. DeAngelis. 1990. The global carbon cycle. American Scientist 78:310-326.

LISA DILLING
Center for Science and Technology Policy Research
Cooperative Institute for Research in Environmental Sciences
University of Colorado
1333 Grandview Ave., 488 UCB
Boulder, CO 80309-0488

EDUCATION

1997	Ph.D.	University of California, Santa Barbara, CA ,Biological Sciences
1989	B.A.	Harvard University, Cambridge, MA, Biology, *magna cum laude*

EMPLOYMENT HISTORY

2004-present	Visiting Fellow, Cooperative Institute for Research in Environmental Sciences, University of Colorado, Boulder CO
2003-present	Project Scientist II, Environmental and Societal Impacts Group, National Center for Atmospheric Research, Boulder CO
2002-2003	Visiting Scientist, Environmental and Societal Impacts Group, National Center for Atmospheric Research, Boulder CO
1999-2002	Co-Chair, Carbon Cycle Interagency Working Group, U.S. Global Change Research Program

1998-2002   Program Manager, Carbon Cycle Program, Office of Global Programs, National
             Oceanic and Atmospheric Administration, Silver Spring, MD.
1997-1998   Associate Program Manager for Ocean-Atmosphere Carbon Exchange Study
             and Atlantic Climate Change Program, Office of Global Programs, National
             Oceanic and Atmospheric Administration (through UCAR), Silver Spring,
             MD.
1996-1997   National Sea Grant Fellow, International Development, Office of Global
             Programs, National Oceanic and Atmospheric Administration, Silver Spring,
             MD

PROFESSIONAL SERVICE/ACTIVITIES

2003-present   North American Carbon Plan (NACP) Implementation Revision Committee
2003-present   Carbon, Climate and Society Initiative, Integrated Graduate Research and
                Traineeship Program (IGERT), University of Colorado, Boulder

PUBLICATIONS

Moser S and Dilling L. 2004. Making Climate Hot: Communicating the urgency and challenge
of global climate change. Environment 46: 32-46.

Dilling L, and MA Brzezinski. 2004. Quantifying marine snow as a food choice for zooplankton
using stable silicon isotope tracers. Journal of Plankton Research 26:1105-1114.

Dilling L, Doney S, Edmonds J, Gurney KR, Harriss R, Schimel D, Stephens B, and Stokes G.
2003. The role of carbon cycle observations and knowledge in carbon management. Annual
Review of Environment and Resources 28:521-58.

Dilling L, and AL Alldredge. 2000. Fragmentation of marine snow by swimming
macrozooplankton: A new process impacting carbon cycling in the sea. Deep Sea Res. I
47:1227-1245.

Dilling L, J Wilson, D Steinberg, and AL Alldredge. 1998. Feeding by the euphausiid *Euphausia
pacifica* and the copepod *Calanus pacificus* on marine snow. Mar. Ecol. Prog. Ser. 170: 189-
201.

Dilling L. and AL Alldredge. 1993. Can chaetognath fecal pellets contribute to carbon flux? Mar.
Ecol. Prog. Ser. 92:51-58.

RECENT PRESENTATIONS

Dilling L. "Toward carbon governance: Challenges for science and policy across scales."
Association of American Geographers. 2005 Annual Meeting.

Dilling L. "In Search of Pasteur's Quadrant: "Use-inspired" Carbon Cycle Science" 2005 Center for Science and Technology Policy Research Symposium.

Dilling L, Pielke Jr, R, and Sarewitz, D. "Pilot study on reconciling supply and demand: Who are the consumers of information on the North American carbon balance?" American Geophysical Union 2004 Fall Meeting.

Dilling L, Doney S, Edmonds J, Gurney K, Harriss R, Schimel D, Stephens B, Stokes G. "A review of the role of carbon cycle science in supporting carbon management policy" American Geophysical Union 2003 Fall Meeting.

Pielke, Jr., Sarewitz D, Dilling L, and Conant R. "Carbon Cycle Science: Reconciling Supply and Demand" North American Carbon Program 2003 PI meeting

AWARDS

CIRES Visiting Fellowship 2004-present
NOAA Cash Award, 1998, 1999, 2000, 2001
Dean John A. Knauss Marine Policy Fellowship, 1995
National Science Foundation Graduate Fellowship, 1991-1993

PROFESSIONAL MEMBERSHIPS

American Geophysical Union

DAVID M. FAIRMAN
Vice President, Consensus Building Institute, Inc.
131 Mt. Auburn Street, Cambridge, MA 02138
Tel. (617) 492-1414 ext. 20

Professional Experience
Feb.1997-present **Consensus Building Institute**   Cambridge, MA
    Vice President (7/99-present)
    Senior Associate (2/97-6/99)
Facilitator, trainer, researcher and manager for non-profit dispute resolution consulting firm. Facilitate negotiations among government, business and civil society stakeholders on economic and social development, environmental protection and natural resource use. Design and teach training courses on negotiation, mediation and consensus-building for public, non-profit and private organizations. Recent and current project conveners include World Bank, Asian Development Bank, U.S. Agency for International Development, U.S. Dept. of Housing and Urban Development, Florida Dept. of Environmental Protection, Council of State Governments, American Cancer Society, United Way of America, Harvard University.

Feb.2000-present **MIT-Harvard Public Disputes Program**   Cambridge, MA

Associate Director
Initiate and direct research projects on application of dispute resolution/consensus building principles and strategies to public policy arenas. Develop strategies and materials for teaching negotiation and dispute resolution skills in secondary, university and professional education settings.

1991-1996   **Private practice** Cambridge, MA
                   Conflict Resolution Consultant
Designed and taught executive training courses on strategies for using negotiation and consensus building to integrate environmental, social and economic objectives in national and international policy-making. Analyzed and recommended strategies for policy integration at the national and international level. Clients included Netherlands Ministry of Housing, Spatial Planning and Environment; UN Commission on Sustainable Development, UN Development Program; U.S. Agency for International Development.

1989-1991   **Endispute, Inc.**   Cambridge, MA
                   Public Policy Mediator
Assessed public policy conflicts at national, state, and local levels; developed and implemented consensus-building and conflict resolution strategies. Managed stakeholder consultation on siting process for low-level radioactive waste facility. Taught negotiation and conflict management skills to public officials. Clients included American Energy Assurance Council; Maine Low-level Radioactive Waste Authority; Massachusetts Dept. of Industrial Accidents; U.S. Army Corps of Engineers.

1989   **Somerville Community Development Corporation**   Somerville, MA
                   Landlord-Tenant Mediator and Counselor

1987-1988   **Harvard College** South Asia *Sheldon Fellow*

Education
1998   ***Massachusetts Institute of Technology*** *Cambridge, MA*
                   Ph.D., Political Science. Dissertation examined negotiation strategies of advocates for natural resource policy reform in developing countries, based on extensive field research on forest policy reform in Philippines and Thailand.

1987   ***Harvard University***   *Cambridge, MA*
                   Bachelor of Arts, summa cum laude in History and Literature.
                   Awards: Phi Beta Kappa, Sheldon Fellowship for postgraduate study, E.C. Cumming Prize for outstanding thesis, History and Literature Prize for academic achievement, Adams House Arms Citation for contributions to residential community.

Professional Affiliations

Alliance for International Conflict Prevention and Resolution. Board member. Chairman, Education and Outreach Committee.

U.S. Environmental Protection Agency: Senior Mediator, ADR Roster
U.S. Institute for Environmental Conflict Resolution: Senior Mediator, Roster of
Conflict Resolution Professionals.
Lincoln Institute of Land Policy: Faculty Associate
Association for Conflict Resolution: Practitioner Member
Council on Foreign Relations: Term Member

SELECTED ASSESSMENT AND FACILITATION PROJECTS
(References available on request)

Asian Development Bank, Chasma Right Bank Irrigation Project Social Assessment. *2001-02.*
*Senior advisor for assessment of unresolved social issues relating to major irrigation
project in Pakistan. Developed assessment strategy with CBI field consultant (Prof. Adil
Najam); reviewed and edited draft assessment report; advised on agenda and work plan
for consultative workshop; provided continuous oversight and advice to CBI field
consultant.*

National Public Housing Assessment Policy Dialogue. *2001-02. Lead facilitator for national
policy dialogue convened by U.S. Department of Housing and Urban Development
(HUD) on public housing assessment. Issues included legal basis for assessment,
assessment criteria and methods, and use of assessment results. In parallel, facilitated
meeting of public housing industry organizations to develop industry proposals on
assessment. Participants include HUD Deputy Assistant Secretaries and staff, four
national housing industry associations, three residents' associations, and technical
analysts from National Academy of Public Administration. Dialogue is ongoing, pending
submission of industry proposals.*

National Energy Policy Initiative. *2001-02. Project manager and co-lead facilitator for
convening and facilitating a national energy issues assessment and an expert workshop,
in conjunction with the Rocky Mountain Institute. Assessment gathered and synthesized
views of 75 leading energy policy stakeholders from business, government, advocacy and
academic institutions. Workshop involved twenty-two of the country's leading energy
policy experts in joint drafting process. Facilitated drafting process to produce 25-page
consensus report and recommendations to inform current Congressional and
Administration development of national energy policy. Gave Congressional testimony
and participated in Congressional briefing and media outreach on the report.*

Florida Department of Environmental Protection Phosphorus Rule Development. *2001. Co-
lead facilitator for rule development process to resolve 12-year controversy over
management of phosphorus run-off from agricultural lands into Everglades Protection
Area. Issues include maximum permissible phosphorus concentration, compliance test
procedures, and permitting/enforcement action to be taken in event of non-compliance.
Participants included Federal EPA and National Parks Service, State DEP, regional
Water Management District, agricultural producer groups, regulated municipalities
regional, state and national environmental groups, and scientific researchers. Process
narrowed range of disagreement on scientific and technical issues.*

PAVE PAWS Upgrade Issues Assessment. 2000. Lead assessor of potential for dialogue and consensus building to resolve conflict over health and safety risks of military radar installation at the Massachusetts Military Reservation; conducted 40 stakeholder interviews, prepared assessment, facilitated public meeting and development of recommendations for further action. Process led to commitment by public agencies to joint health effects study.

World Bank Forest Policy Evaluation Workshop. 1999-2000. Advised on development of agenda, participation guidelines and ground rules; co-led facilitation of 2-day workshop event; and drafted post-workshop report for global workshop to review World Bank's Evaluation of its forest policy. Issues included balance among environmental, economic and social goals in current policy, and impacts of policy implementation in over 100 countries worldwide. Participants included World Bank staff, donor and borrower governments, forest conservation advocacy groups, commercial timber companies and forest researchers. Participants reached consensus on numerous strengths and weaknesses of Evaluation report, and on recommendations for further action by the World Bank and other forest policy stakeholders.

SELECTED REPORTS AND PUBLICATIONS

Reframing the Forest: The Politics of Tropical Forest Policy Reform. Washington, D.C.: Resources for the Future Press, forthcoming 2003.

"Integrating Conflict Resolution into the High School Curriculum: The Example of Workable Peace." Co-author with Stacie Nicole Smith. In N. Noddings, ed., Educating for Global Citizenship: Challenges and Opportunities. *New York: Teachers College Press, 2003.*

"Fulfilling the Promise of Environmental Conflict Resolution." Co-author with Lisa Bingham, Dan Fiorino, and Rosemary O'Leary." In L. Bingham and R. O'Leary, eds., Evaluating Environmental and Public Policy Dispute Resolution Programs. Washington, D.C.: Resources for the Future Press, forthcoming 2003.

Consensus Building and Conflict Resolution Toolkit for National Standard Setting Processes. (IKEA-WWF Cooperation for Forest Stewardship, 2002. Available at http://www.piec.org/pathfinder/pages/instruments.html.

National Energy Policy Initiative: Expert Group Report. Snowmass, CO: Rocky Mountain Institute, March 2002.Available at www.nepinitiative.org.

Juan F. Consent Decree Issues Assessment. Confidential report to the Connecticut Department of Children and Families, Juan F. Next Friends (child welfare plaintiffs) and the Office of the Court-Appointed Monitor. January 2001.

Convening Report for Proposed PAVE PAWS Stakeholder Working Group. Cambridge, MA: Consensus Building Institute, March 2000.

"Producing Consensus." Co-author with Sarah McKearnan. In *The Consensus Building Handbook*, L.Susskind et al., eds. Thousand Oaks, CA: Sage Publications, 1999.

Reforming Natural Resource Policies in Developing Countries: The Politics of Forest Policy Reform in the Philippines, Thailand and Costa Rica, 1980-1996. Cambridge, MA: MIT Department of Political Science (dissertation), 1998.

Alternative Dispute Resolution Practitioners Guide. Co-author with Scott Brown and Christine Cervenak. Washington, D.C.: United States Agency for International Development, 1997.

"The Global Environment Facility: Haunted by the Shadow of the Future," In Robert Keohane and Marc Levy, eds., *Institutions for Environmental Aid: Pitfalls and Promise.* Cambridge, MA: MIT Press, 1996.

"Old Fads, New Lessons: Learning from Economic Development Assistance." Co-author with Michael Ross. In Robert Keohane and Marc Levy, eds., *Institutions for Environmental Aid: Pitfalls and Promise.* Cambridge, MA: MIT Press, 1996.

Richard A. Houghton
The Woods Hole Research Center
P.O. Box 296
Woods Hole, Massachusetts 02543
Tel: (508) 540-9900; Fax: (508) 540-9700

Education

1965	B.A.	Biology, Hamilton College
1969		Long Island University, Special student in Oceanography
1979	Ph.D.	Ecology, S.U.N.Y., Stony Brook

Employment History

1989-present	Senior Scientist, Woods Hole Research Center, Woods Hole, Massachusetts
1993-1994	Visiting Senior Scientist, Office of Mission to Planet Earth, NASA, Wash., D.C.
1987-1989	Associate Scientist, Woods Hole Research Center, Woods Hole, MA
1984-1987	Associate Scientist, Ecosystems Center, Marine Biological Laboratory, Woods Hole, MA
1975-1984	Res. Assoc., Ecosystems Center, Marine Biological Laboratory, Woods Hole, MA
1967-1974	Research Associate, Biology Department, Brookhaven National Lab., Upton, NY

Professional Service/Activities

Marquis Who's Who in America
Honorary Doctorate from the Faculty of Forest Science, University of Munich
Associate Editor, *Environmental Reviews*
National Technical Advisory Committee for NIGEC
Landsat Pathfinder Science Working Group
Member - Ecological Society of America
Member - American Geophysical Union
Member - Sigma Xi

Publications

Houghton, R.A. 1999. The annual net flux of carbon to the atmosphere from changes in land use 1850-1990. Tellus 51B:298-313.

Houghton, R.A., J.L. Hackler and K.T. Lawrence. 1999. The U.S. carbon budget: contributions from land-use change. Science 285:574-578.

Houghton, R.A. 2000. Emissions of carbon from land-use change. Pages 63-76 *in: The Carbon Cycle* (T.M.L. Wigley and D.S. Schimel, editors), Cambridge University Press, New York, NY.

Houghton, R.A. and J.L. Hackler. 2000. Changes in terrestrial carbon storage in the United States. 1. The roles of agriculture and forestry. Global Ecology and Biogeography 9:125-144.

Houghton, R.A., J.L. Hackler and K.T. Lawrence. 2000. Changes in terrestrial carbon storage in the United States. 2. The role of fire and fire management. Global Ecology and Biogeography 9:145-170.

Noble, I., M. Apps, R. Houghton, D. Lashof, W. Makundi, D. Murdiyarso, B. Murray, W. Sombroek and R. Valentini. 2000. Implications of different definitions and generic issues. Pages 53-126 in: R.T. Watson, I.R. Noble, B. Bolin, N.H. Ravindranath, D.J. Verardo and D.J. Dokken (editors). Land Use, Land-Use Change, and Forestry. A Special Report of the IPCC. Cambridge University Press, New York.

Houghton, R.A. and J.L. Hackler. 2001. Carbon Flux to the Atmosphere from Land-Use Changes: 1850 to 1990. ORNL/CDIAC-131, NDP-050/R1. Carbon Dioxide Information Analysis Center, U.S. Department of Energy, Oak Ridge National Laboratory, Oak Ridge, Tennessee, U.S.A.

Pacala, S.W., G.C. Hurtt, D. Baker, P. Peylin, R.A. Houghton, R.A. Birdsey, L. Heath, E.T. Sundquist, R.F. Stallard, P. Ciais, P. Moorcroft, J.P. Caspersen, E. Shevliakova, B. Moore, G. Kohlmaier, E. Holland, M. Gloor, M.E. Harmon, S.-M. Fan, J.L. Sarmiento, C.L. Goodale, D. Schimel and C.B. Field. 2001. Consistent land- and atmosphere-based U.S. carbon sink estimates. Science 292:2316-2320.

Schimel, D.S., J.I. House, K.A. Hibbard, P. Bousquet, P. Ciais, P. Peylin, B.H. Braswell, M.J. Apps, D. Baker, A. Bondeau, J. Canadell, G. Churkina, W. Cramer, A.S. Denning, C.B. Field, P. Friedlingstein, C. Goodale, M. Heimann, R.A. Houghton, J.M. Melillo, B. Moore III, D. Murdiyarso, I. Noble, S.W. Pacala, I.C. Prentice, M.R. Raupach, P.J. Rayner, R.J. Scholes, W.L. Steffen and C. Wirth. 2001. Recent patterns and mechanisms of carbon exchange by terrestrial ecosystems. Nature 414:169-172.

DeFries, R.S., R.A. Houghton, M.C. Hansen, C.B. Field, D. Skole and J. Townshend. 2002. Carbon emissions from tropical deforestation and regrowth based on satellite observations for the 1980s and 90s. Proceedings of the National Academy of Sciences 99:14256-14261.

Goodale, C.L., M.J. Apps, R.A. Birdsey, C.B. Field, L.S. Heath, R.A. Houghton, J.C. Jenkins, G. H. Kohlmaier, W. Kurz, S. Liu, G.-J. Nabuurs, S. Nilsson and A.Z. Shvidenko. 2002. Forest carbon sinks in the northern hemisphere. Ecological Applications 12:891-899.

Houghton, R.A. 2002. Magnitude, distribution and causes of terrestrial carbon sinks and some implications for policy. Climate Policy 2:71-88.

Hurtt, G.C., S.W. Pacala, P.R. Moorcroft, J. Caspersen, E. Shevliakova, R.A. Houghton and B. Moore III. 2002. Projecting the future of the U.S. carbon sink. Proceedings of the National Academy of Sciences 99:1389-1394.

Houghton, R.A. 2003. Revised estimates of the annual net flux of carbon to the atmosphere from changes in land use and land management 1850-2000. Tellus 55B:378-390.

Houghton, R.A. 2003. Why are estimates of the terrestrial carbon balance so different? Global Change Biology 9:500-509.

Houghton, R.A. and J.L. Hackler. 2003. Sources and sinks of carbon from land-use change in China. Global Biogeochemical Cycles 17(2):1034.

House, J.I., I.C. Prentice, N. Ramankutty, R.A. Houghton and M. Heimann. 2003. Reconciling apparent inconsistencies in estimates of terrestrial CO_2 sources and sinks. Tellus 55B:345-363.

Gregg Marland
Environmental Sciences Division
Oak Ridge National Laboratory
P.O. Box 2008
Oak Ridge, TN 37831-6335
Tel: (865) 241-4850; Fax: (865) 574-2232

Education

1964	B.S.	Virginia Polytechnic Institute, Blacksburg, VA
1964-1966		Washington University, St. Louis, MO
1972	Ph.D.	University of Minnesota, Minneapolis, MN

Employment History

2000-present	Distinguished Scientist, Oak Ridge National Laboratory
1987-2000	Senior Staff Scientist, Oak Ridge National Laboratory
1975-1987	Staff Scientist, Institute for Energy Analysis, Oak Ridge Associated Universities
1970-1975	Assistant Professor of Geology, Indiana State University

Professional Service/Activities

Committee on Global Change Research - National Research Council

Lead author - IPCC (Intergovernmental Panel on Climate Change): Special Report on Carbon Capture and Storage

Lead author - IPCC: Third Assessment Report, Land-use Change and Forestry

Lead author - IPCC: Special Report on Land Use, Land-Use Change and Forestry

Lead author - IPCC: Second Assessment Report, Energy Primer

Publications

Marland, G., A Brenkert and J. Olivier. 1999. CO_2 from fossil fuel burning: a comparison of ORNL and EDGAR estimates of national emissions. Environmental Science and Policy 2:265-273.

Marland, G. and B. Schlamadinger. 1999. The Kyoto Protocol could make a difference for optimal forest-based CO_2 mitigation strategy: some results from GORCAM. Environmental Science and Policy 2:111-124.

Schlamadinger B. and G. Marland. 1999. Net effect of forest harvest on CO_2 emissions to the atmosphere: a sensitivity analysis on the influence of time. Tellus 51B:314-325.

Andres, R.J., D.J. Fielding, G. Marland, T.A. Boden and N. Kumar. 1999. Carbon dioxide emissions from fossil-fuel use, 1751-1950. Tellus 51B:759-765.

Sampson, R.N., R.J. Scholes, et al. 2000. Additional human-induced activities – Article 3.4, In Land use, land-use change, and forestry, A special report of the Intergovernmental Panel on Climate Change, R.T. Watson, I.R. Noble, B. Bolin, N.H. Ravindranath, D.J. Verardo and D.J. Dokken (eds.), Cambridge University Press, UK, pp. 181-281.

Marland, G., B. Schlamadinger and R. Matthews. 2000. "Kyoto Forests" and a broader perspective on management. Science 290:1895-1896.

Kheshgi, H., R. Prince and G. Marland. 2000. The potential of biomass fuels in the context of global climate change: focus on transportation fuels. Annual reviews of Energy and Environment 25:1999-2444.

Marland, G., K. Fruit and R. Sedjo. 2001. Accounting for sequestered carbon: the question of permanence. Environmental Science and Policy 4:259-268.

Marland, G., T.O. West and J. Fenderson. 2001. Carbon emitted, carbon saved; CDIAC Communications Newsletter, Issue no. 28, Carbon Dioxide Information Analysis Center, Oak ridge National Laboratory, Oak Ridge, TN.

Marland, G., B.A. McCarl and U. Schneider. 2001. Soil carbon: policy and economics. Climatic Change 51:101-117.

West, T.O. and G. Marland. 2002. A synthesis of carbon sequestration, carbon emissions, and net carbon flux in agriculture: comparing tillage practices in the United States. Agricultural Ecosystems and Environment 91:217-232.

West, T.O. and G. Marland. 2002. Net carbon flux from agricultural ecosystems: methodology for full carbon cycle analyses. Environmental Pollution 116:439-444.

Marland, G. and T. Boden. 2002. The increasing concentration of atmospheric CO_2: how much, when, and why? In Proceedings of the International seminar on nuclear war and planetary emergencies 26th session, R. Ragaini (ed.), 19-24 August, 2001, Erice, Italy, World Scientific Publishing Co., River Edge, New Jersey, USA, pp. 283-295.

Pielke, R.A. Sr., G. Marland, R.A. Betts, T.N. Chase, J.L. Eastman, J.O. Niles, D.S. Niyogi and S.W. Running. 2002. The influence of land-use change and landscape dynamics on the climate system - relevance to climate change policy beyond the radioactive effect of greenhouse gases. Philosophical Transactions of the Royal Society of London A. 360:1705-1719.

Schlamadinger, B., L. Aukland, S. Berg, D. Bradley, L. Ciccarese, V. Dameron, A. Faaij, M. Jackson, G. Marland and R. Sikkema. 2002. Forest-based carbon mitigation projects; options for carbon accounting and for dealing with non-permanence, United Nations Framework

Convention on Climate Change, FCCC/WEB/2002/12,4 Sept.2002, http://unfccc.int/resources/webdocs/2002/12.pdf.

Marland, E. and G. Marland. 2003. The treatment of long-lived, carbon-containing products in inventories of carbon dioxide emissions to the atmosphere. Environmental Science and Policy 6:139-152.

Huston, M.A. and G. Marland. 2003. Carbon management and biodiversity. J. of Environmental Management 67:77-86.

Marland, G., R.A. Pielke Sr., M. Apps, R. Avissar, R.A. Betts, K.J. Davis, P.C. Frumhoff, S.T. Jackson, L. Joyce, P. Kauppi, J. Katzenberger, K.G. MacDicken, R. Neilson, J.O. Niles, D.D.S. Niyogi, R.J. Norby, N. Pena, N. Sampson and Y. Xue. 2003. The climatic impacts of land surface change and carbon management, and the implications for climate-change mitigation policy. Climate Policy 3:149-157.

Marland, G., T.O. West, B. Schlamadinger and L. Canella. 2003. Managing soil organic carbon in agriculture: the net effect on greenhouse gas emissions. Tellus 55B:613-621.

West, T.O. and G. Marland. 2003. Net carbon flux from agriculture: carbon emissions, carbon sequestration, crop yield, and land-use change. Biogeochemistry 63:73-82.

Marland, G., C.T. Garten Jr., W.M. Post and T.O. West. 2003. CSiTE studies on carbon sequestration in soils. Energy – The International Journal (in press).

Sedjo, R.A. and G. Marland. 2003. Inter-trading permanent emissions credits and rented temporary carbon emissions offsets: some issues and alternatives. Climate Policy (in press).

West, T.O., G. Marland, W.M. Post, A.W. King, A.K. Jain and K. Andrasko. 2003. Carbon management response curves: estimates of temporal carbon dynamics. Environmental Management (in press).

Marland, G., D. Archer, G. Benford, M. Ishikawa, F.B. Metting, F.M. Orr Jr. and T. Volk. 2003. Biological Options toward stabilization of greenhouse gas concentrations in the Earth's atmosphere. Aspen Global Change Institute (in press).

ADAM ZACHARY ROSE

The Pennsylvania State University

213 Walker Building, University Park, PA 16802

Phone: (814) 863-0179 Fax: (814) 863-7943

EDUCATION

Ph.D. (Economics)	Cornell University	1974
M.A. (Economics)	Cornell University	1972
B.A. (Economics)	University of Utah	1970

RESEARCH AND TEACHING FIELDS

Environmental & Resource Economics	Natural & Man-Made Hazards
Energy Economics	Economic Development
Regional & Urban Economics	Applied General Equilibrium Analysis

EMPLOYMENT HISTORY

Professor, Department of Geography, The Pennsylvania State University	2002-
Professor, Department of Energy, Environmental, and Mineral Economics, The Pennsylvania State University (Department Head, 1988-02)	1988-02
Professor, Department of Mineral Resource Economics, West Virginia University (Department Chairman, 1986-88)	1984-88
Associate Professor, Department of Mineral Resource Economics, West Virginia University (Department Chairman, 1981-83)	1981-84
Faculty Associate, Regional Research Institute, West Virginia University	1981-88
Assistant Professor, Department of Economics, University of California, Riverside	1977-81
Lecturer, Department of Economics, University of California, Riverside	1975-77
Senior Council Economist, New York State Council of Economic Advisers	1974-75

RECENT VISITING POSITIONS

Visiting Fellow, East-West Center	2004
Resident Visitor, Resources for the Future	2001

RECENT ADVISORY POSITIONS

U.S. EPA Advisory Panel on the Second Generation Climate Policy Model	2004
Chair, NSF Site Review Team, Center for Decision-Making under Uncertainty	2004
National Academy of Sciences, Panel on Economic Benefits of Improved Seismic Monitoring	2003-
Consortium for Atlantic Regional Assessment of Climate Change, Advisory Council	2003-
NSF/Earthquake Engineering Research Institute, Panel on Research Opportunities for Earthquake Engineering	2001-03
Pennsylvania Consortium for Interdisciplinary Environmental Policy, Committee on Climate Change and Energy (Chair, 2001-03)	2001-
Multidisciplinary Center for Earthquake Engineering Research, Research Committee	2001
Editorial Board, *Energy Policy*	2000-
Editorial Board, *Pacific and Asian Journal of Energy*	1995-
Editorial Board, *Resource and Energy Economics*	1993-

SELECTED PUBLICATIONS

Recent Refereed Journal Articles
"Modeling Regional Economic Resilience to Disasters: A Computable General Equilibrium Analysis of Water Service Disruptions," Journal of Regional Science, forthcoming (with S. Liao).

"Reducing the Conflict Between Climate Policy and Energy Policy in the U.S.: The Important Role of the States," Energy Policy, forthcoming (with T. Peterson).

"Incentive-Based Approaches to Greenhouse Gas Mitigation in Pennsylvania: Protecting the Environment and Promoting Fiscal Reform," Widener Law Journal, forthcoming (with R. McKinstry and C. Ripp).

"A Greenhouse Gas Emissions Inventory for Pennsylvania," Journal of the Air and Waste Management Association, forthcoming (with B. Yarnal and others).

"Global Climate Change and the Value of Solar Energy in the U.S. Agriculture," Land Economics, forthcoming (with R. Kamat and J. Shortle).

"Defining and Measuring Economic Resilience to Disasters," Disaster Prevention and Management, Vol. 13, No. 4, 2004, pp. 307-14.

"Interregional Burden-Sharing of Greenhouse Gas Mitigation in the United States," Mitigation and Adaptation Strategies for Global Change, Vol. 9, No. 3, 2004, pp. 477-500 (with Z. Zhang).

"Greenhouse Gas Mitigation Action Planning," Penn State Environmental Law Review, Vol. 12, No. 1, 2003, pp. 153-71.

"A Dynamic Analysis of the Marketable Permits Approach to Global Warming Policy: A Comparison of Spatial and Temporal Flexibility," Journal of Environmental Economics and Management, Vol. 44, No. 1, 2002, pp. 45-69 (with B. K. Stevens).

"Business Interruption Losses from Natural Hazards: Conceptual and Methodological Issues in the Case of the Northridge Earthquake," Environmental Hazards: Human and Policy Dimensions, Vol. 4, No. 2, 2002, pp. 1-14 (with D. Lim).

"Greenhouse Gas Reduction in the U.S.: Identifying Winners and Losers in an Expanded Permit Trading System," Energy Journal, Vol. 23, No. 1, 2002, pp. 1-18 (with G. Oladosu).

"An Economic Analysis of Flexible Permit Trading in the Kyoto Protocol," International Environmental Agreements, Vol. 1, No. 2, 2001, pp. 219-42 (with B. K. Stevens).

"Characterizing Economic Impacts and Responses to Climate Change," Global and Planetary Change, Vol. 25, No. 2, 2000, pp. 67-81 (with J. Shortle and others).

Recent Research Reports

Greenhouse Gas Emissions Inventory for Pennsylvania, Report to the Pennsylvania Department of Environmental Protection, Center for Integrated Regional Assessment, The Pennsylvania State University, 2003 (with B. Yarnal and others).

User Costs in Seismic Risk Management for Urban Infrastructure Systems, Report to the National Science Foundation, Department of Geography, University of Washington, 2002 (with S. Chang and others).

Chad-Cameroon Development Project: Economic Impact Assessment of Cameroon, Report to the World Bank for ExxonMobil, URS Corporation, Houston, TX, 2002 (with F. Bayne).

Mid-Atlantic Regional Assessment (MARA): The Impacts of Climate Change, Report to the U.S. Environmental Protection Agency, The Pennsylvania State University, 2000 (with A. Fisher and others).

Recent Contributions to Public Documents

National Research Council. 2005. Economic Benefits of Improved Seismic Monitoring, Washington, DC: National Academy Press, 2004 (with other members of a National Academy of Sciences Panel).

European Union. 2003. "Understanding Sources of Resiliency to Natural Hazards," in A. van der Veen et al. (eds.) <u>Proceedings of the Joint NEDEIS and University of Twente Workshop: In Search of a Common Methodology for Damage Estimation</u>, Bruxelles: Office for Official Publications of the European Communities, 2003, pp. 137-50 (with S. Liao).

National Institute of Building Sciences/Federal Emergency Management Agency, "Indirect Economic Losses," <u>Flood Loss Estimation Methodology</u>, Washington, DC, 2003 (with H. Cochrane and S. Chang).

Earthquake Engineering Research Institute, <u>Securing Society Against Catastrophic Loss: A Research and Technology Transfer Plan</u>, Report to the National Science Foundation, Oakland, CA, 2002 (with other members of an Expert Review Panel).

PROFESSIONAL PRESENTATIONS (Selected)

Conferences of Professional Organizations

American Economic Association Meetings: 1986, 1987

American Association for the Advancement of Science Meetings: 1991, 1992, 1994

American Society of Civil Engineers
  Structural Engineers Joint World Congress: 1998
  U.S. Conference on Lifeline Earthquake Engineering: 1999

Association of American Geographers Meetings: 2003, 2004

Association of Environmental and Resource Economists
  European Meetings: 1992, 1993, 1996, 1997, 2000, 2001, 2003, 2004
  World Congress: 1998

International Association for Energy Economics
  International Meetings: 1999, 2001, 2002
  North American Meetings: 2000

International Society for Ecological Economics Biennial Meetings: 1996, 1998

Regional Science Association
  European Meetings: 1994
  North American Meetings: 1990, 1992, 1994-97, 1999, 2001-03
  Pacific Meetings: 1995

Western Economic Association, 1980, 1994, 1999

PROFESSIONAL RESEARCH ACTIVITIES (Recent and Current)

Major Grant and Contract Research

– Principal Investigator and Project Director, Pennsylvania Department of Environmental Protection contract – Economic Impact Modeling of Pennsylvania's Indigenous Resources, 2004-.

– Track A Team Leader, National Institute of Building Sciences/Federal Emergency Management contract – Independent Study to Assess Future Savings from Hazard Mitigation Activities (requested by U.S. Congress), 2003-04 (consultant to Applied Technology Council).

– Principal Investigator and Project Director, National Science Foundation grant (through the Multidisciplinary Center for Earthquake Engineering Research) – Modeling Regional Economic Losses from Earthquakes: LA Lifeline Study, 2003-04.

- Co-Principal Investigator and Project Leader, Pennsylvania Department of Environmental Protection contract – Pennsylvania Greenhouse Gas Emissions Inventory, 2001-03.
- Co-Principal Investigator, U.S. Department of Energy NIGEC contract – Climate Change and Policy Impacts on the Southeastern U.S. Economy, 2000-01 (subcontractor through University of Alabama; renewed Phase 2, 2001-02).

Consultantships
- DHS Center for Risk and Economic Analysis of Terrorism Events – Analyzing Threats to the Economy through Computable General Equilibrium Analysis, 2004-.
- U.S. Department of Defense – Independent Review Panel on Economic Impact Analysis Methodology for the Base Realignment and Closure 2005 Process (through Booz Allen Hamilton), 2004.
- U.S. Department of Homeland Security – Development of a Framework to Estimate the Economic Impacts of Terrorist Attacks (subcontractor to ABS Consulting), 2004.
- ICF Consulting, Inc. – Upgrading the Outer Continental Shelf Economic Impact Models for the Gulf of Mexico and the Alaska OCS, 2003-.
- Center for Energy and Economic Development – Economic Impact of Wind-Generated Electricity Displacement of Coal, 2003.

Thomas J. Wilbanks
Oak Ridge National Laboratory
P.O. Box 2008
Oak Ridge, TN 37831-6206
Tel: (865) 574-5515; Fax: (865) 576-2943

Education

1960	B.A.	Trinity University
1967	M.A.	Syracuse University
1969	Ph.D.	Syracuse University

Research Interests

Realizing sustainable development.

Relationships between society and technology.

Responses to concerns about global environmental change.

Energy and environmental policy analysis, including technology assessment; regional assessment; environmental, social, and economic impact assessment; R&D policy.

Institution-building, especially for R&D activities and for energy and environmental policymaking and decisionmaking.

Geographic scale as an issue in sustainability science, including roles of cross-scale interactions.

Regional development, particularly problems of developing regions and cross-cultural comparisons of determinants.

Employment History

1987-present	Corporate Research Fellow and Leader, Global Change and Developing Country Programs, Oak Ridge National Laboratory
1998-present	Associate, Belfer Center for Science and International Affairs, Harvard University
200-2002	Acting Co-Director, Oak Ridge Center for Advanced Studies
1983-present	Adjunct Professor of Geography, University of Tennessee
1980-1987	Associate Director and Head of Programs and Planning, Energy Division, Oak Ridge National Laboratory
1980-1981	Acting Director, Energy Division, Oak Ridge National Laboratory
1977-1980	Senior Planner, Energy Division, Oak Ridge National Laboratory
1974-1977	Research Fellow, Science and Public Policy Program, The University of Oklahoma
1973-1977	Associate Professor and Chair, Department of Geography, The University of Oklahoma
1973	Research Director, Syracuse-Yugoslav Project on Environmental Policy and Planning, Ljubljana, Yugoslavia
1971-1972	Executive Director, Urban Transportation Institute, Syracuse University
1969-1973	Assistant Professor of Geography, Syracuse University
1969	Lecturer in Geography, Syracuse University

Professional Service/Activities

Member - Science Steering Group, U.S. Carbon Cycle Program

Co-author - scaling chapter of conceptual framework report, Millennium Ecosystem Assessment, UN Environment Programme et al.

Member - IPCC Working Group II (Impacts, Adaptation, and Vulnerabilities), Third Assessment Report; lead author of chapter 7 (human settlements, energy, and industry) and of the synthesis report and summary for policymakers

Member - Advisory Committee, Human-Environmental Research Observatories, NSF-supported national program led by the Pennsylvania State University

Coordinator - Inter-regional Forum, U.S. National Assessment of Vulnerabilities to Climate Variability and Change

Member - Board on Earth Sciences and Resources, National Academy of Sciences/National Research Council

Member - Committee on Human Dimensions of Global Change, National Academy of Sciences/National Research Council

Member - Panel on Public Participation in Environmental Assessment and Decision Making, National Academy of Sciences/National Research Council

Publications

Wilbanks, T.J. 1002. Geography and Technology, in Technology and Geography: A Social History, S. Brunn, S. Cutter, J. Harrington (eds.), Dordrecht: Kluwer.

Wilbanks, T.J., et al. 2003. Possible Responses to Global Climate Change: Integrating Mitigation and Adaptation. Environment 45(5):28-38.

Wilbanks, T.J. and R. Kates. 2003. Making the Global Local: Responding to Climate Change Concerns from the Bottom Up. Environment 45(3):12-23.

Wilbanks, T.J., and D. Capistrano, et al. 2003. Dealing with Scale, Conceptual Framework, Millennium Ecosystem Assessment, Kuala Lumpur, Island Press, pp. 107-126.

Wilbanks, T.J. and E.A. Parson, et al. 2003. Understanding Climatic Impacts, Vulnerabilities, and Adaptation in the United States: Building a Capacity for Assessment. Climatic Change 57:9-42.

Wilbanks, T.J., S. Cutter, and D. Richardson. 2003. The Geographical Dimensions of Terrorism, Routledge, New York.

Wilbanks, T.J., R. Kates, and R. Abler. 2003. Global Change and Local Places: Estimating, Understanding, and Reducing Greenhouse Gases, Cambridge University Press.

Wilbanks, T.J. 2003. Geographic Scaling Issues in Integrated Assessments of Climate Change, in Scaling Issues in Integrated Assessment, J. Rotmans and D. Rothman (eds.). Swets and Zeitlinger 5-34.

Wilbanks, T.J., and W.C. Clark, et al. 2000. Assessing Vulnerability to Global Environmental Risks, Discussion Paper 2000-12, Environment and Natural Resources Program, Kennedy School of Government, Harvard University.

Wilbanks, T.J., A. Wolfe, and N. Kerchner. 2001. Public Involvement on a Regional Scale. Environmental Assessment Review 21:431-448.

Wilbanks, T.J., and P. Stern. 2001. Policy Implications and Needs for Further Knowledge, New Tools for Environmental Protection: Education, Information, and Voluntary Measures, National Academy of Sciences/National Research Council.

Wilbanks, T.J. and R.W. Kates. 1999. Global Change in Local Places. Climatic Change 43(3):601-628.

Wilbanks, T.J. 1994. Sustainable Development' in Geographic Context. Annals, Association of American Geographers, 84:541-57.

Wilbanks, T.J. 1992. Energy Policy Responses to Concerns about Global Climate Change, in Global Climate Change: Implications, Challenges and Mitigation Measures, S. Majumdar, et al. (eds.), Pennsylvania Academy of Sciences, Easton, PA, pp. 452-70.

Wilbanks, T.J., et al. 1989. Decision Making, in Energy Technology R&D: What Could Make a Difference?, W. Fulkerson et al. (eds.), ORNL-6541, Vol. 2, Oak Ridge National Laboratory, pp. 123-37.

Wilbanks, T.J. 1988. Impacts of Energy Development and Use, 1888-2088, in Earth '88: Changing Geographic Perspectives, National Geographic Society, Washington, pp. 96-114.

Wilbanks, T.J. 1985. Geography and National Policy, Annals, Association of American Geographers, LXXV, pp. 4-10.

Wilbanks, T.J., and R. Lee. 1985. Policy Analysis in Theory and Practice, in Large-Scale Energy Projects: Assessment of Regional Consequences, T.R. Lakshmanan and B. Johansson (eds.), North-Holland, Amsterdam 273-303.

Wilbanks, T.J., and E. Aronson, et al. 1984. Energy Use: The Human Dimension, W.H. Freeman, San Francisco.

Wilbanks, T.J. 1982. Is Comprehensive Analysis of Critical Interactions Possible?, in Energy, Economics, and the Environment, G. Daneke (ed.), D.C. Heath, Lexington, MA, pp. 91-110.

Wilbanks, T.J. 1980. Location and Well-being, Harper and Row, New York, 462 pp.

Wilbanks, T.J., and D.E. Kash, et al. 1976. Our Energy Future: The Role of Research, Development, and Demonstration in Reaching a National Consensus on Energy Supply, University of Oklahoma Press, Norman, 482 pp.

Wilbanks, T.J., and D.E. Kash, et al.. 1974. A Methodology and Documentation for Consistent Analysis of Energy Alternatives, Science and Public Policy Program, University of Oklahoma, Norman, Vol 4., 1400 pp.

Gregory P. Zimmerman
Environmental Sciences Division
Oak Ridge National Laboratory
P.O. Box 2008
Oak Ridge, TN 37831-6200
Tel: (865) 574-5815; Fax: (865) 574-5788

Education

1977   M.S. Mechanical Engineering, University of Tennessee, Knoxville
1975   B.S. Mechanical Engineering, University of Tennessee, Knoxville

Employment History

1977-present   Research Staff Member, Oak Ridge National Laboratory

Publications

Zimmerman, G.P. 2001. Project leader for U.S. Nuclear Regulatory Commission, Final Environmental Impact Statement for the Construction and Operation of an Independent Spent Fuel Storage Installation on the Reservation of the Skull Valley Band of Goshute Indians and the Related Transportation Facility in Tooele County, Utah (Volumes 1 and 2), NUREG-1714, U.S. Nuclear Regulatory Commission, Office of Nuclear Material Safety and Safeguards,Washington, D.C., December 2001.

Berry, J.B., C.J. Coomer, R.C. DeVault, M.R. Hilliard, P.J. Hughes, M.P. Ternes and G.P. Zimmerman. 2000. Case Studies in Sustaining DoD Readiness, 26th Environmental Symposium and Exhibition, March 27 to 30, 2000, Long Beach, Calif.; National Defense Industrial Association, Arlington, Va., Report No. P00-106353.

G. Ostrouchov, G.P. Zimmerman, J.J. Beauchamp, V.V. Fedorov and D.J. Downing. 1999. Evaluation of Statistical Methodologies Used in U.S. Army Ordnance and Explosives Work, ORNL/TM-13588, Oak Ridge National Laboratory, Oak Ridge, Tenn., September 1999.

Zimmerman, G.P. 1996. Technical Core Team Leader for U.S. Department of Energy, Performance Evaluation of the Technical Capabilities of DOE Sites for Disposal of Mixed Low-Level Waste, DOE/ID-10521 (Vols. 1, 2, and 3) and SAND96-0721 (Vols. 1, 2, and 3), prepared by Sandia National Laboratories, Albuquerque, New Mexico, March 1996.

J.D. Tauxe, D.W. Lee, J.C. Wang and G.P. Zimmerman. 1995. A Comparative Subsurface Transport Analysis for Radioactive Waste Disposal at Various DOE Sites, P95-79881,

Proceedings of the 1995 Fall Meeting of the American Geophysical Union, San Francisco, Calif., December 11-15, 1995.

G.P. Zimmerman. 1994. Coal Technology Characterization and Discharges, Appendix A in Estimating Externalities of Coal Fuel Cycles; Report Number 3 on the External Costs and Benefits of Fuel Cycles: A Study by the U.S. Department of Energy and the Commission of the European Communities, prepared by the Oak Ridge National Laboratory and Resources for the Future; McGraw-Hill, September 1994.

ATTACHMENT 3. BIOGRAPHIES OF SOCCR CHAPTER AUTHORS
(I.E., SAP 2.2 "LEAD CHAPTER AUTHORS")

Richard A. Birdsey
USDA Forest Service
11 Campus Blvd. Ste. 200
Newtown Square, PA 19073
Tel: 610-557-4091; Fax: 610-557-4095

Education

1971	B.S. Rensselaer Polytechnic Institute – Anthropology
1975	M.S. State University of New York (Syracuse) – World Forestry
1989	Ph.D. State University of New York (Syracuse) – Forest Management

Research Interests

Quantitative methods for large-scale ecosystem and watershed inventories, methods to estimate national carbon budgets from forest inventory data, estimates of historical U.S. forest carbon sources and sinks, accounting rules and guidelines for U.S. forests, forest management strategies to increase carbon sequestration, assessments of U.S. forest resources, impacts of multiple stresses on forests, adaptation to climate change.

Employment History

1991 – Present	Program Manager, Global Change Research, USDA Forest Service
1989-1991	Staff Scientist, Forest Inventory and Analysis, USDA Forest Service
1979-1989	Research Forester, Forest Inventory and Analysis, USDA Forest Service
1976-1979	Forester, U.S. Peace Corps (Ecuador)

Publications

Alexeyev, V.; Birdsey, R.; Stakanov, V.; Korotkov, I. 1995. Carbon in vegetation of Russian forests: methods to estimate storage and geographical distribution. Water, Air, and Soil Pollution 82:271-282.

Birdsey, R.A. 1996. Carbon storage for major forest types and regions in the conterminous United States. In: Forests and Global Change Volume Two - Forest Management Opportunities. ed. by R. Neil Sampson and Dwight Hair. Washington, DC: American Forests. Pp. 1-25 plus appendices.

Birdsey, Richard A. 2003. Current and historical trends in use, management, and disturbance of U.S. forestlands. In: Kimble, J.M. et al. (Eds.), The Potential of U.S. forest soils to sequester carbon and mitigate the greenhouse effect. Boca Raton, FL: CRC Press. Pp. 15-34.

Birdsey, Richard A. 2004. Data gaps for monitoring forest carbon in the United States: an inventory perspective. In: Mickler, Robert A., eds. Environmental Management. 33(Supplement 1): S1-S8.

Birdsey, R.A. and L.S. Heath. 2001. Forest inventory data, models, and assumptions for monitoring carbon flux. In: SSSA Special Publication no. 57, Soil Carbon Sequestration and the Greenhouse Effect. Madison, WI: Soil Science Society of America. Pp 125-135.

Birdsey, R. A.; A. J. Plantinga; L. S. Heath. 1993. Past and prospective carbon storage in United States forests. Forest Ecology and Management. 58:33-39.

Birdsey, Richard A.; Heath, L.S. 1995. Carbon changes in U.S. forests. In: Productivity of America's Forests and Climate Change ed. by Linda A. Joyce. Ft. Collins, CO: USDA Forest Service, Gen. Tech. Report RM-271. 70p.

Casperson, John P.; Pacala, Stephen W.; Jenkins, Jennifer C.; Hurtt, George C.; Moorcraft, Paul R.; Birdsey, Richard A. 2000. Contributions of land-use history to carbon accumulation in U.S. forests. Science 290: 1148-1151.

Jenkins, J.; Birdsey, R; Pan, Y. 2000. Biomass and NPP estimation for the mid-Atlantic region (USA) using plot-level forest inventory data. Ecological Applications 11(4): 1174-1193.

Jenkins, Jennifer C.; Chojnacky, David C.; Heath, Linda S.; Birdsey, Richard A. 2003. National-scale biomass estimators for United States tree species. Forest Science 49(1): 12-35.

Mickler, Robert A.; Birdsey, Richard A.; Hom, John. (Eds.) 2000. Responses of Northern U.S. forests to environmental change. Ecological Studies 139. Springer-Verlag, New York. 578 p.

Pacala, S.W., G.C. Hurtt, D.Baker, P.Peylin, R.A. Houghton, R.A. Birdsey, et al. 2001. Consistent land- and atmosphere-based U.S. carbon sink estimates. 2001. Science 292: 2316-2320.

Pan, Yude; Hom, John; Birdsey, Richard; McCullough, Kevin. 2004. Impacts of rising nitrogen deposition on N exports from forests to surface waters in the Chesapeake Bay Watershed. Environmental Management. 33: S120-S131.

Scott D. Bridgham
Center for Ecology and Evolutionary Biology
5289 University of Oregon
Eugene, OR 97403-5289
Tel: (541) 346-1466; Fax: (541) 346-2364

Education

1991 Ph.D. School of the Environment, Duke University, Durham, NC
1986 M.S. Department of Ecology, Evolution and Behavior, University of Minnesota,
    Minneapolis, MN.,
1982 B.A. University of Maine, Orono, with Highest Honors
1980 B.A. University of Maine, Orono, with Highest Honors

Research Interests

Carbon and nutrient cycling, wetland ecology, trace gas production, climate change,
biogeochemistry, microbial ecology, plant community structure, plant-nutrient interactions,
invasion ecology, restoration.

Employment History

Associate Professor, Center for Ecology and Environmental Biology and Environmental Studies
    Program, University of Oregon, 2003 - present.
Associate Professor, Department of Biological Sciences, University of Notre Dame, 2001 - 2002.
Assistant Professor, Department of Biological Sciences, University of Notre Dame, 1994 –2001.
Research Associate, Natural Resources Research Institute, University of Minnesota, Duluth,
    1992 - 1994.
Postdoctoral Research Associate, Natural Resources Research Institute, University of Minnesota,
    Duluth, 1991 - 1992. Advisors: Carol Johnston and John Pastor.
Research Assistant, School of the Environment, Duke University, 1986 - 1991.
Research and Teaching Assistant, Department of Ecology, Evolution and Behavioral Biology,
    University of Minnesota, 1983 - 1986.
Field Research Technician, USDA Forest Service, Orono, ME, 1978 -1979.

Professional Service/Activities

Milton Ellis Award for Academic Distinction in English--1980, University of Maine.
Eugene A. Jordan Memorial Scholarship for Outstanding Academic Achievement in Zoology-
    1982, University of Maine.
National Science Foundation Grant for Improving Doctoral Dissertation Research--1988 - 1991.
        Department of Energy Global Change Distinguished Postdoctoral Fellowship--Sept. 1991
        - Aug. 1993.
National Science Foundation CAREER Award, 9/96 - 8/2001.
Editorial Board of *Soil Science Society of America Journal*, 1994-1997.
Editorial Board of *Wetlands*, 1997-2000.
Chair of the Division S-10, Wetland Soils, of the Soil Science Society of America, 2001-2002.
Editorial Board of *Biogeochemistry*, 2004-current.

Publications

Bridgham, S. D. and C. J. Richardson. 1992. Mechanisms controlling soil respiration (CO_2 and CH_4) in southern peatlands. Soil Biology and Biochemistry 24:1089-1099.

Bridgham, S. D., C. A. Johnston, J. Pastor, and K. Updegraff. 1995. Potential feedbacks of northern wetlands on climate change. BioScience 45:262-274.

Bridgham, S. D., K. Updegraff, and J. Pastor. 1998. Carbon, nitrogen, and phosphorus mineralization in northern wetlands. Ecology 79:1545-1561.

Updegraff, K., S. D. Bridgham, J. Pastor, and P. Weishampel. 1998. Hysteresis in the temperature response of carbon dioxide and methane production in peat soils. Biogeochemistry 43:253-272.

Bridgham, S. D., J. Pastor, K. Updegraff, T. J. Malterer, K. Johnson, C. Harth, and J. Chen. 1999. Ecosystem control over temperature and energy flux in northern peatlands. Ecological Applications 9: 1345-1358.

Weltzin, J. F., J. Pastor, C. Harth, S. D. Bridgham, K. Updegraff, and C. T. Chapin. 2000. Response of bog and fen plant communities to warming and water-table manipulations. Ecology 81: 3464-3478.

Updegraff, K., S. D. Bridgham, J. Pastor, P. Weishampel, and C. Harth. 2001. Response of CO_2 and CH_4 emissions in peatlands to warming and water-table manipulation. Ecological Applications 11: 311-326.

Weltzin, J. F., S. D. Bridgham, J. Pastor, J. Chen, and C. Harth. 2003. Potential effects of warming and drying on peatland plant community composition. Global Change Biology 9:1-11.

Pastor, J., J. Solin, S. D. Bridgham, K. Updegraff, C. Harth, P. Weishampel, and B. Dewey. 2003. Global warming and the export of dissolved organic carbon from boreal peatlands. Oikos 100: 380-386.

Vile, M. A., S. D. Bridgham, R. K. Wieder, and M. Novák. 2003. Atmospheric sulfur deposition alters pathways of gaseous carbon production in peatlands. Global Biogeochemical Cycles 17:1058-1064.

Vile, M. A., S. D. Bridgham, and R. K. Wieder. 2003. Response of anaerobic carbon mineralization rates to sulfate amendments in a boreal peatland. Ecological Applications 13:720-734.

Bridgham, S. D., and C. J. Richardson. 2003. Endogenous versus exogenous nutrient control over decomposition in North Carolina peatlands. Biogeochemistry 65:151-178.

Keller, J. K., J. R. White, S. D. Bridgham, and J. Pastor. 2004. Climate change effects on carbon and nitrogen mineralization in peatlands through changes in soil quality. Global Change Biology 10:1053-1064.

Keller, J. K., S. D. Bridgham, C. T. Chapin, C. M. Iversen. 2005. Limited effects of six years of fertilization on carbon mineralization dynamics in a Minnesota fen. Soil Biology and Biochemistry 37(6):1197-1204.

Pendall, E., S. Bridgham, P. J. Hanson, B. Hungate, D. W. Kicklighter, D. W. Johnson, B. E. Law, Y. Luo, J. P. Megonigal, M. Olsrud[1], M. G. Ryan, and S. Wan. *In Press*. Belowground process responses to elevated CO_2 and temperature: a discussion of observations, measurement methods, and models. New Phytologist.

Ken Caldeira
Energy and Environmental Sciences Directorate
Lawrence Livermore National Laboratory
7000 East Ave, L-103
Livermore, CA 94550 USA
Tel: (925) 423-4191; Fax: (925) 422-6388

Education

1991   Ph.D.   New York University, Atmospheric Sciences, Department of Applied Science
1988   M.S.   New York University, Atmospheric Sciences, Department of Applied Science
1978   B.A.   Rutgers College, Philosophy

Research Interests

Long-term evolution of climate and geochemical cycles; ocean carbon sequestration; numerical simulation of climate, carbon, and biogeochemistry; marine biogeochemical cycles; approaches to supplying energy services with diminished environmental footprint

Employment History

1995- present Physicist/Environmental Scientist (LLNL) Research ocean carbon cycle, atmospheric CO_2, ocean/sea-ice physics, climate, and energy systems

1993-1995   Post-Doctoral Researcher (LLNL) Research the ocean carbon cycle, atmospheric CO_2 and climate

1991-1993   NSF Earth Sciences Postdoctoral Fellow (Earth Systems Science Center & Dept. of Geosciences, The Pennsylvania State University) Role of the carbonate-silicate cycle in long-term atmospheric CO_2 content and climate

Publications

Govindasamy, B., S. Thompson, A. Mirin, M. Wickett, K. Caldeira and C. Delire Increase of carbon cycle feedback with climate sensitivity: results from a coupled climate and carbon cycle model Tellus B 57 (2) 153 -- 163 DOI: 10.1111/j.1600-0889.2005.00135.x, 2005.

Caldeira, K., G. Morgan, D. Baldocchi, P. Brewer, C.-T. Arthur Chen, G.-J. Nabuurs, N. Nakicenovic, P. Robertson, A portfolio of carbon management options, in Towards CO2 Stabilization: Issues, Strategies, and Consequences, A SCOPE report (Scientific Committee on Problems in the Environment) C. Field and M. Raupach, editors (pp. 103-129, Island Press, Washington DC) 2004.

Thompson SL, Govindasamy B, Mirin A, K. Caldeira, C. Delire, J. Milovich, M. Wickett, and D. Erickson. Quantifying the effects of CO2-fertilized vegetation on future global climate and carbon dynamics. Geophysical Research Letters 31 (23): Art. No. L23211 DEC 11 2004.

Hoffert, M.I., and K. Caldeira, Climate change and energy — An overview, Encyclopedia of Energy, C. Cleveland, ed., Academic Press, San Diego, CA, 359-380, 2004.

Caldeira, K., and M.E. Wickett, Anthropogenic carbon and ocean pH, Nature 425, 365-365, 2003.

Caldeira, K., A.K. Jain, and M.I. Hoffert, Climate sensitivity uncertainty and the need for energy without CO2 emission, Science 299, 2052-2054, 2003.

Herzog, H., K. Caldeira, and J. Reilly. An issue of permanence: Assessing the effectiveness of ocean carbon sequestration, Climatic Change 59, 293–310, 2003.

Ridgwell, A,J,, M.J. Kennedy, and K. Caldeira, Carbonate deposition, climate stability, and Neoproterozoic ice ages, Science 302, 859–862, 2003.

Caldeira, K., M.E. Wickett, and P.B. Duffy. Depth, radiocarbon and the effectiveness of direct CO2 injection as an ocean carbon sequestration strategy. Geophysical Research Letters, 10.1029/2001GL014234, 2002.

Govindasamy, B., S. Thompson, P. Duffy, K. Caldeira and C. Delire, Impact of geoengineering schemes on the terrestrial biosphere. Geophysical Research Letters, 10.1029/2002GL015911, 2002.

Hoffert, M.I., K. Caldeira, G. Benford, D.R. Criswell, C. Green, H. Herzog, J.W. Katzenberger, H.S. Kheshgi, K.S. Lackner, J.S. Lewis, W. Manheimer, J.C. Mankins, G. Marland, M.E. Mauel, L.J. Perkins, M.E. Schlesinger, T. Volk, and T.M.L. Wigley, Advanced technology paths to global climate stability: Energy for a greenhouse planet, Science 295, 981–987, 2002.

Lutz, M., R.L. Dunbar, and K. Caldeira, Regional variability in the vertical flux of particulate organic carbon in the ocean interior, Global Biogeogeochemical Cycles 16, U91-U110, 2002.

Govindasamy, B., P.B. Duffy, and K. Caldeira, Land use changes and Northern Hemisphere cooling, Geophysical Research Letters 28, 291-294, 2001.

Herzog, H., K. Caldeira and E. Adams, Carbon Sequestration via Direct Injection. In J H Steele, S A Thorpe and K K Turekian (eds) Encyclopedia of Ocean Sciences Vol. 1, pp 408 - 414. London, UK: Academic Press, 2001.

Caldeira, K., and P.B. Duffy, The role of the Southern Ocean in uptake and storage of anthropogenic carbon dioxide, Science 287, 620–622, 2000.

Caldeira, K., and G.H. Rau, Accelerating carbonate dissolution to sequester carbon dioxide in the ocean: Geochemical implications, Geophysical Research Letters, 27, 225–228, 2000.

Govindasamy, B., and K. Caldeira, Geoengineering Earth's radiation balance to mitigate CO2-induced climate change, Geophysical Research Letters 27, 2141-2144, 2000.

Guilderson, T.P., K. Caldeira, and P.B. Duffy, Radiocarbon as a diagnostic tracer in ocean and carbon cycle modeling, Global Biogeochemical Cycles, 14, 887–902, 2000.

Caldeira, K., and Berner, R., Seawater pH and atmospheric carbon dioxide (Technical comment), Science 286, 2043a–2043a, 1999.

Rau, G.H., and Caldeira, K. Enhanced carbonate dissolution: A means of sequestering waste CO2 as ocean bicarbonate. Energy Conversion and Management 40, 1803–1813, 1999.

Kerrick, D.M. and K. Caldeira, Metamorphic CO2 degassing from orogenic belts, Chemical Geology 145 213–232, 1998.

Hoffert M.I., K. Caldeira, A.K. Jain, E.F. Haites, L.D.D. Harvey, S. D. Potter, M.E. Schlesinger, S. H. Schneider, R.G. Watts, T. M. L Wigley, and D. J. Wuebbles. Energy implications of future stabilization of atmospheric CO2 content. Nature 395, 881–884, 1998.

Francisco P. Chavez
Monterey Bay Aquarium Research Institute (MBARI)
7700 Sandholdt Road
Moss Landing, CA 95039-9644
Tel: (831) 775-1709; Fax: (831) 775-1620

Education

1987   Ph.D.   Botany, Duke University
1977   B.S.   Oceanography, Humboldt State University

Research Interests

Biology and chemistry of the ocean in relation to natural climate variability and global change. Global carbon cycle. Instrumentation and systems for long-term ocean observing. Satellite remote sensing.

Employment History

2000-present   Senior Scientist , MBARI
2000-present   Faculty (courtesy), Stanford University
1996-2000   Associate Scientist (III), MBARI
1992-1996   Associate Scientist (II), MBARI
1990-present   Research Associate, University of California, Santa Cruz
1987-1992   Assistant Scientist, MBARI

Professional Service/Activities

Member - JGOFS time series oversight committee
Reviewer - Chilean Oceanographic Program, Peruvian Fisheries Program
NSF Alan Waterman award committee
NSF Advisory Committee for the Geoscience Directorate
Board of Directors - Center for Integrated Marine Technologies
Science Team - Global Eulerian Observations

Publications

Barber, R.T. and F.P. Chavez. 1983. Biological consequences of El Niño. Science 222:1203-1210.
Chavez, F.P., R.T. Barber and H. Soldi S. 1984. Propagated temperature changes during onset and recovery of the 1982-83 El Niño. Nature 309:47-49.
Barber, R.T. and F.P. Chavez. 1986. Ocean variability in relation to living resources during the 1982-83 El Niño. Nature 319:279-285.
Chavez, F.P. (1987). El Niño y la Oscilacion del Sur. Investigacion y Ciencia (Spanish edition of Scientific American) 128:46-55.

Martin, J.H. et al. 1994. Testing the iron hypothesis in ecosystems of the equatorial Pacific Ocean. Nature 371:123-129.

Paytan, A., M. Kastner and F.P. Chavez. 1996. Glacial to interglacial fluctuations in productivity in the Equatorial Pacific as indicated by marine barite. Science 274:1355-1377.

Coale, K.H et al. 1996. A massive phytoplankton bloom induced by an ecosystem-scale iron fertilization experiment in the equatorial Pacific Ocean. Nature 383:495-501.

Johnson, K.S., F.P. Chavez and G.E. Friederich. 1999. Continental shelf sediment as a primary source of iron for coastal phytoplankton. Nature 398:697-700.

Chavez, F.P., P.G. Strutton, G.E. Friederich, R.A. Feely, G.A. Feldman, D. Foley and M.J. McPhaden. 1999. Biological and chemical response of the equatorial Pacific Ocean to the 1997-1998 El Niño. Science 286:2126-2131.

Chavez, F.P., J.P. Ryan, S. Lluch-Cota and M. Ñiquen C. 2003. From anchovies to sardines and back-Multidecadal change in the Pacific Ocean. Science 299:217-221.

Chavez, F.P. and J.R. Toggweiler. 1995. Physical estimates of global new production: the upwelling contribution, In Upwelling in the Ocean: Modern Processes and Ancient Records, Summerhayes, C.P., Emeis, K.C., Angel, M.V., Smith, R.L., and Zeitzschel, B., (eds.), p. 313-320, J. Wiley & Sons, Chichester.

Chavez, F.P., J.T. Pennington, R. Herlien, H. Jannasch, G. Thurmond and G.E. Friederich. 1997. Moorings and drifters for real-time interdisciplinary oceanography. Journal of Atmospheric and Oceanic Technology 14:1199-1211.

Chavez, F.P. and C. Collins, eds. 1998. Studies of the California Current System, Deep-Sea Research II, Volume 45.

Olivieri, R.O. and F.P. Chavez. 2000. A model of plankton dynamics for the coastal upwelling system of Monterey Bay, California. Deep-Sea Research II 47:1077-1105.

Pennington, J.T. and F.P. Chavez. 2000. Seasonal fluctuations of temperature, salinity, nitrate, chlorophyll and primary production at station H3/M1 over 1989-1996 in Monterey Bay, California. Deep-Sea Research II 47:947-973.

Chavez, F.P. and C. Collins, eds. 2000. Studies of the California Current System Part 2, Deep-Sea Research II 47:5-6.

Johnson, K.S., F.P. Chavez, V.A. Elrod, S.E. Fitzwater, J.T. Pennington, K.R. Buck and P.M. Walz. 2001. The annual cycle of iron and the biological response in central California coastal waters. Geophysical Research Letters 28:1247-1250.

Johnson, K.S., C.K. Paull, J.P. Barry and F.P. Chavez. 2001. A decadal record of underflows from a coastal river into the deep sea. Geology 29:1019-1022.

Friederich, G., P. Walz, M. Burczynski and F.P. Chavez. 2002. Inorganic Carbon in the Central California Upwelling System During the 1997-1999 El Niño -La Nina Event. Progress in Oceanography 54:185-204.

Chavez, F.P, C.A. Collins, A. Huyer and D. Mackas (eds). 2002. El Niño along the west coast of North America. Progress in Oceanography 54:1-6.

Collins, C.A. J.T. Pennington, C.G. Castro, T.A. Rago and F.P. Chavez. 2003. The California Current system off Monterey, California: Physical and biological coupling. Deep-Sea Research II. doi:10.1016/S0967-0645(03)00134-6

Richard T. Conant
Natural Resource Ecology Laboratory
Colorado State University
Campus Delivery 1499
Fort Collins, CO 80523-1499
Telephone: (970) 491-1919 ; FAX: (970) 491-1965

Education

May 1997   Ph.D.   Arizona State University, Botany (Ecology).
            Dissertation: *Carbon pools and fluxes along a semiarid gradient in Northern
            Arizona.*
Dec 1990   B.A.   University of Colorado, Environmental Biology
            Undergraduate Research: *Effects of nutrient amendments on plant nutrient
            concentrations in two Alpine Tundra communities.*

Professional Services/Activities

Journal reviews (last 2 years only): Canadian Journal of Soil Science, Chemosphere, Ecology,
    Ecological Applications, Ecological Modelling, Environmental Management, Environmental
    Pollution, Frontiers in Ecology and the Environment, Global and Planetary Change, Global
    Change Biology, Journal of Arid Environments, Pedosphere, Soil & Tillage Research, Soil
    Biology & Biochemistry, Soil Science Society of America Journal.
Proposal reviews (last 2 years only): Kearney Foundation, Science Foundation Ireland, National
    Science Foundation-Ecosystems, Department of Energy-NIGEC, NASA-Earth System
    Fellowship Panel.
Contributing Author, IPCC Good Practice Guidance for Land Use, Land-use Change, and
    Forestry, 2004.
Contributing Author, IPCC Special Report on Land Use, Land Use Change, and Forestry, 2000.
Chapter Secretary, Rocky Mountain Chapter, Ecological Society of America 1999-2000.
Co-convener, The Contribution of Terrestrial and Anthropogenic Processes to Atmospheric CO_2
    Concentrations in the Mid Continent NACP Intensive, Des Moines, IA, September 15-17,
    2004.
Co-convener, Reconciling supply and demand of carbon cycle science, Fort Collins, CO,
    September 18-19, 2004.
Co-convener, Taking the PULSE of Colorado's Front Range, Denver, CO, November 17, 2004.

Employment History

2002- Research Sci. II Natural Resource Ecology Lab Colorado St. Univ.
2001- Faculty Affiliate Dept. of Forest, Range, and Colorado St. Univ.
                          Watershed Stewardship
2001 - Faculty Affiliate Grad. Degree Program in Ecology Colorado St. Univ.
2000-2002 Scientist Natural Resource Ecology Lab Colorado St. Univ.

Publications

Conant, R. T., J. Six, and K. Paustian. 2004. Land use effects on soil carbon fractions in the southeastern United States: II. Change in soil carbon fractions along a forest to pasture chronosequence. Biology and Fertility of Soils **40**:194-200.

Nabuurs, G.-J., et al (including R. T. Conant). 2004. LUCF-sector good practice guidance. *in* J. Penman, M. Gytarsky, T. Hirishi, T. Krug, and D. Kruger, editors. IPCC Good Practice Guidance for LULUCF. Institute for Global Environmental Strategies, Hayama, Japan.

Conant, R. T., and K. Paustian. 2004. Grassland management activity data: Current sources and future needs. Environmental Management **33**:467-473.

Ogle, S. M., R. T. Conant, and K. Paustian. 2004. Deriving grassland management factors for a carbon accounting method developed by the intergovernmental panel on climate change. Environmental Management **33**:474-484.

Conant, R. T., J. Six, and K. Paustian. 2003. Land use effects on soil carbon fractions in the southeastern United States: I. Management intensive versus extensive grazing. Biology and Fertility of Soils **36**:386-392.

Pielke, R. A. and R. T. Conant. 2003. Best practices in prediction for decision making: Lessons from the atmospheric and earth sciences. Ecology 84:1351-1358.

Conant, R. T. 2003. Grazer-dominated ecosystems. *in* Encyclopedia of Life Sciences. Nature Publishing Group, London.

Conant, R. T. and K. Paustian. 2002. Potential soil carbon sequestration in overgrazed grassland ecosystems. Global Biogeochemical Cycles **16**: art. no. 1143

Conant, R. T., K. Paustian, and E. T. Elliott. 2002. Pasture land use in the southeastern US: Implications for C sequestration. Pages 423-432 *in* R. Lal, R. F. Follet, and J. M. Kimble, editors. Agricultural practices and policies for carbon sequestration in soil. CRC Press, Boca Raton.

Conant, R. T., Paustian, K., and E. T. Elliott. 2001. Changes in soil carbon from improved grassland management. Ecological Applications **11**:343-355.

Sampson, R. M., et al (including R. T. Conant). 2000. Additional human-induced activities - article 3.4. Pages 180-281 *in* R. T. Watson, I. R. Noble, B. Bolin, N. H. Ravindranath, D. J. Verardo, and D. J. Dokken, editors. Land Use, Land-Use Change, and Forestry. Camridge University Press, Cambridge.

Kenneth J. Davis
Department of Meteorology
The Pennsylvania State University
512 Walker Building
University Park, PA
Tel: (814) 863-8601; Fax: (814) 865 3663

Education

1987   A. B.   Princeton University
1992   Ph.D.   University of Colorado, Astrophysical, Planetary and Atmospheric Sciences
1994   Postdoc National Center for Atmospheric Research,Trace gas micrometeorology

Employment History

2000-present   Associate Professor, Dept of Meteorology, The Pennsylvania State University
1996-2000   Assistant Professor, Department of Soil, Water, and Climate, U. of Minnesota.
Fall 1996   Guest Scientist, Institute for Atmospheric Physics, German Aerospace Research Establishment (DLR).
1995-1996   Research Associate, University of Colorado, Cooperative Institute for Research in Environmental Sciences
1995-1996   Visiting Scientist, Mesoscale and Microscale Meteorology Division, National Center for Atmospheric Research
1993-1994   Postdoctoral Fellow, NCAR, Advanced Studies Program.
1989-1992   NASA Graduate Student Researchers Program Fellow, APAS Department, University of Colorado.
1989-1992   Graduate Research Assistant, Advanced Studies Program, NCAR.

Relevant Publications

Desai, A.R., Bolstad, P., Cook, B.D., Davis, K.J. and Carey, E.V. 2005. Comparing net ecosystem exchange of carbon dioxide between an old-growth and mature forest in the upper Midwest, USA, *Agricultural and Forest Meteorology* **128**(1-2): 33-55.

Cook, B.D., K.J. Davis, W. Wang, A. Desai, B.W. Berger, R. M. Teclaw, J. G. Martin, P.V. Bolstad, P.S. Bakwin, C. Yi and W. Heilman, 2004. Carbon exchange and venting anomalies in an upland deciduous forest in northern Wisconsin, USA. *Agricultural and Forest Meteorology,* **126**, 271-295.

Werner, C., K. J. Davis, P. S. Bakwin, C. Yi, D. Hurst, and L. Lock, 2003. Interannual variability of methane exchange over a temperate-boreal lowland and wetland forest. *Global Change Biology*, **9**, 1251-1261.

Davis, K.J., P.S. Bakwin, B.W. Berger, C. Yi, C. Zhao, R.M. Teclaw and J.G. Isebrands, 2003. The annual cycle of CO_2 and H_2O exchange over a northern mixed forest as observed from a very tall tower. *Global Change Biology*, **9**, 1278-1293.

Berger, B.W., K.J. Davis, P.S. Bakwin, C. Yi and C. Zhao, 2001. Long-term carbon dioxide fluxes from a very tall tower in a northern forest: Flux measurement methodology. *J. Atmos. Oceanic Tech.*, **18**, 529-542.

Additional Publications

Yi, C., K. J. Davis, P. S. Bakwin, A.S. Denning, N. Zhang, A. Desai, J.Ch.-H. Lin, and C. Gerbig, 2004, The observed covariance between ecosystem carbon exchange and atmospheric boundary layer dynamics in North Wisconsin, *Journal of Geophysical Research*, **109**(D08302): doi10.1029/2003JD004164.

Hurwitz, M.D., D.M. Ricciuto, K.J. Davis, W. Wang, C. Yi, M.P. Butler, P.S. Bakwin. Advection of carbon dioxide in the presence of storm systems over a northern Wisconsin forest, 2004. *J. Atmos. Sci.*, **61**, 607-618.

Denning, A.S., M. Nicholls, L. Prihodko, I. Baker, P.-L. Vidale, K.J. Davis, and P.S. Bakwin' 2003. Simulated and observed variations in atmospheric CO_2 over a Wisconsin forest. *Global Change Biology*, **9**, 1241-1250.

Baker, I., A.S. Denning, N. Hanan, L. Prihodko, M. Uliasz, P.-L. Vidale, K.J. Davis, and P.S. Bakwin, 2003. Simulated and Observed Fluxes of Sensible and Latent Heat and CO_2 at the WLEF-TV Tower Using SiB2.5. *Global Change Biology*, **9**, 1262-1277.

MacKay, D.S., D.E. Ahl, B.E. Ewers, S.T. Gower, S.N. Burrows, S. Samanta and K.J. Davis, 2002. Effects of aggregated classifications of forest composition on estimates of evapotranspiration in a northern Wisconsin forest. *Global Change Biology*, **8**, 1253-1266.

Christopher B. Field
Department of Global Ecology
Carnegie Institution
Stanford, CA 94305

Education

| 1975 | A.B., Biology | Harvard College |
| 1981 | Ph.D., Biology | Stanford University |

Employment History

1981 -- 1984	Assistant Professor, Biology, University of Utah
1984 -- 2002	Staff Scientist, Carnegie Institution of Washington
1986 -- 1989	Assistant Professor by courtesy, Stanford University
1989 -- 1996	Associate Professor by courtesy, Stanford University
1996 -- 2005	Professor by courtesy, Stanford University
2002 -- present	Director, Department of Global Ecology, Carnegie Institution
2005 -- present	Professor of Biological Sciences, Stanford University

Professional Service/Activities

Member, US National Academy of Sciences (2001)

Ecological Society of America Aldo Leopold Fellow (2000)

Senior Fellow by courtesy, Stanford Institute of International Studies (2005)

Service on a variety of National Research Council committees – Currently on Board on Environmental Science and Toxicology (1999)

Service on several editorial boards – Currently on board of Proceedings of the National Academy of Sciences (2001)

Publications

Field, CB, and MR Raupach, editors. 2004. The Global Carbon Cycle: Integrating Humans, Climate, and the Natural World. Island Press, Washington.

Field, CB, and J Kaduk. 2004. The carbon balance of an old-growth forest: Building across approaches. **Ecosystems** 7:525-533.

Field, CB, MR Raupach, and R Victoria. 2004. The global carbon cycle: Integrating humans, climate, and the natural world. Pages 1-13 in CB Field and MR Raupach, editors. The global carbon cycle: Integrating humans, climate, and the natural world. Island Press, Washington, DC.

Gruber, N, P Friedlingstein, CB Field, R Valentini, M Heimann, JE Richey, P Romero-Lankao, E-D Schulze, and C-T A Chen. 2004. The vulnerability of the carbon cycle in the 21st century: An assessment of carbon-climate-human interactions. Pages 45-76 in CB Field and MR Raupach, editors. The Global Carbon Cycle: Integrating Humans, Climate, and the Natural World. Island Press, Washington, DC.

Hayhoe, K, D Cayan, CB Field, PC Frumhoff, EP Maurer, NL Miller, SC Moser, SH Schneider, KN Cahill, EE Cleland, L Dale, R Drapek, RM Hanemann, LS Kalkstein, J Lenihan, CK Lunch, RP Neilson, SC Sheridan, and JH Verville. 2004. Emissions pathways, climate change, and impacts on California. **Proceedings of the National Academy of Sciences of the United States of America** 101:12422-12427.

Horz, H-P, A Barbrook, CB Field, and BJM Bohannan. 2004. The response of soil bacteria to simulated global change. **Proceedings of the National Academy of Sciences of the United States of America** 101: 15136-15141

Melillo, JM, CB Field, and B Moldan, editors. 2003. Interactions of the Major Biogeochemical Cycles: Global Change and Human Impacts. Island Press, Washington, D.C.

Cowling, SA, and CB Field. 2003. Environmental control of leaf area production: Implications for vegetation and land-surface modeling - art. No. 1007. **Global Biogeochemical Cycles** 17:1007.

Hungate, B, J Dukes, M Shaw, Y Luo, and CB Field. 2003. Nitrogen and climate change. **Science** 302:1512-1513.

Melillo, JR, CB Field, and B Moldan. 2003. Element interactions and the cycles of life: An overview. Pages 1-14 *in* JM Melillo, CB Field, and B Moldan, editors. Interactions of the Major Biogeochemical Cycles: Global Change and Human Impacts. Island Press, Washington, D.C.

Zavaleta, E, B Thomas, N Chiariello, G Asner, M Shaw, and C Field. 2003. Plants reverse warming effect on ecosystem water balance. **Proceedings of The National Academy of Sciences of the United States of America** 100:9892-9893.

Zavaleta, E, M Shaw, N Chiariello, B Thomas, E Cleland, C Field, and H Mooney. 2003. Grassland responses to three years of elevated temperature, CO_2, precipitation, and n deposition. **Ecological Monographs** 73:585-604.

DeFries, RS, RA Houghton, MC Hansen, CB Field, D Skole, and J Townshend. 2002. Carbon emissions from tropical deforestation and regrowth based on satellite observations for the 1980s and 1990s. **Proceedings of the National Academy of Sciences** 99:14256-14261.

Goodale, CL, MJ Apps, RA Birdsey, CB Field, LS Heath, RA Houghton, JC Jenkins, GH Kohlmaier, W Kurz, SR Liu, GJ Nabuurs, S Nilsson, and AZ Shvidenko. 2002. Forest carbon sinks in the northern hemisphere. **Ecological Applications** 12:891-899.

Hicke, JA, GP Asner, JT Randerson, CJ Tucker, SO Los, RA Birdsey, JC Jenkins, C Field, and E Holland. 2002. Satellite-derived increases in net primary productivity across north america 1982-1998. **Geophysical Research Letters** 29.

Shaw, MR, ES Zavaleta, NR Chiariello, EE Cleland, HA Mooney, and CB Field. 2002. Grassland responses to global environmental changes suppressed by elevated CO_2. **Science** 298:1987-1990.

Behrenfeld, MJ, JT Randerson, CR McClain, GC Feldman, SO Los, CJ Tucker, PG Falkowski, CB Field, R Frouin, WE Esaias, DD Kolber, and NH Pollack. 2001. Biospheric primary production during an enso transition. **Science** 291:2594-2597.

Hu, S, FS Chapin, MK Firestone, CB Field, and NR Chiariello. 2001. Nitrogen limitation of microbial decomposition in a grassland under elevated CO_2. **Nature (London)** 409:188-191.

Pacala, SW, GC Hurtt, D Baker, P Peylin, RA Houghton, RA Birdsey, L Heath, ET Sundquist, RF Stallard, P Ciais, P Moorcroft, JP Caspersen, E Shevliakova, B Moore, G Kohlmaier, E Holland, M Gloor, ME Harmon, SM Fan, JL Sarmiento, CL Goodale, D Schimel, and

CB Field. 2001. Consistent land- and atmosphere-based us carbon sink estimates. **Science** 292:2316-2319.

David L. Greene
Oak Ridge National Laboratory
National Transportation Research Center
2360 Cherahala Boulevard
Knoxville, Tennessee 37932
Tel. (865) 946-1310

EDUCATION

THE JOHNS HOPKINS UNIVERSITY, Ph.D.; Geography and Environmental Engineering, 1973–78

University of Oregon, M.A.; 1972–73

Columbia University, B.A., 1967–71

EMPLOYMENT HISTORY

OAK RIDGE NATIONAL LABORATORY (ORNL), 1977–PRESENT

1999–Present   Corporate Fellow, Oak Ridge National Laboratory

1989–1999   Senior Research Staff Member II and Manager of Energy Policy Research Programs, Center for Transportation Analysis

1988–1989   Senior Research Analyst, Office of Policy Integration, U.S. Department of Energy (On assignment from ORNL)

1987–1988   Head, Transportation Research Section

1984–1987   Senior Research Staff Member I

1982–1984   Research Staff Member

1980–1982   Leader, Transportation Energy Group

1977–1980   Research Associate

PROFESSIONAL ACTIVITIES

Editor-in-Chief, *Journal of Transportation and Statistics*, 1997–2000

Editorial Board Member, *Journal of Transportation and Statistics*, 2001–2004

Editorial Board Member, *Energy Policy*, 2001–present

Editorial Board Member, Macmillan Encyclopedia of Energy, 1998–2001

Editorial Board Member, *Transportation Quarterly*, 1999–present

Editorial Advisory Board, *Transportation Research A*, 1986–1997

Editorial Advisory Board, *Transportation Research Part D*, 1996–present

National Research Council

> ***Standing Committees:***
>> Chairman, Committee on Energy Conservation and Transportation Demand, A1F01, 1983–1986, 1986–1990; Member, 1993–1998
>> Chairman, Subcommittee on Forecasting Transportation Energy Demand, A1F01(2), 1982–1983

Chairman, Section F, Energy and Environmental Concerns, 1990–1992
Member, Committee on Alternative Fuels, A1F05, 1993–present
Secretary, Task Force on Freight Transportation Data, A1B51, 1989–1996
Member, Committee on Transportation Information Systems and Data Requirements, 1983–1986, 1986–1989

Ad Hoc Committees:

Committee on State Practices in Setting Mobile Source Emissions Standards, 2004–2005
Chair, Committee on Integrating Sustainability into Transportation Planning, 2003–2004
Committee on Effectiveness and Impacts of Corporate Average Fuel Economy (CAFE) Standards, 2001
Committee for the Study of the Impacts of Highway Capacity Improvements on Air Quality and Energy Consumption, 1993–present
Committee on Fuel Economy of Automobiles and Light Trucks, Energy Engineering Board, Commission on Engineering and Technical Systems, 1991–1992
Committee for the Study of High-Speed Surface Transportation in the United States, 1990
Planning Group on Strategic Issues in Domestic Freight Transportation, 1990
Steering Committee for Conference on Transportation, Urban Form, and the Environment, 1990
National Cooperative Highway Research Program, Panel on "Evaluating Alternative Methods of Highway Finance," 1991–1992
Co-Chairman, Conference on Transportation and Energy, Asilomar, California, 1993
Chairman of Organizing Committee, Conference on Transportation and Global Climate Change, Asilomar, California, 1991

Intergovernmental Panel on Climate Change
Lead Author, Working Group III, Fourth Assessment Report, in progress
Lead Author, Working Group III, Third Assessment, 2001
Principal Lead Author, Working Group II, Second Assessment Report, 1995

Association of American Geographers
Board of Directors, Transportation Specialty Group, 1989–1991
Secretary-Treasurer, Transportation Geography Specialty Group, 1980–1982
Editor, *Transportation Geography Newsletter*, 1980–1982

Society of Automotive Engineers, member, 1985–present
Consultant, Eno Transportation Foundation (nonprofit), 1991–1996
Consultant, Transportation Research Board, 1996–1997
International Association for Energy Economics, member

PUBLICATIONS

Books:

D.L. Greene, D.W. Jones and Mark Delucchi, eds., *The Full Costs and Benefits of Transportation*, Springer-Verlag, Heidelberg, 1997.

Transportation and Energy, Eno Foundation for Transportation, Lansdowne, Virginia, 1996.

D.L. Greene and D.J. Santini, eds., *Transportation and Global Climate Change*, American Council for an Energy Efficient Economy, Washington, DC, 1993.

Articles in Professional Journals:

Sheffield, J., et al., "Energy Options for the Future," Journal of Fusion Energy, vol. 23, no. 2, pp. 63-109.

D.L. Greene and J.L.Hopson and J. Li, "Running Out of and Into Oil: Analyzing Global Depletion and Transition Through 2050, *Transportation Research Record 1880*, pp. 1-9, Transportation Research Board, Washington, DC, 2005.

D.L. Greene and P.D. Patterson, M. Sing and J. Li, "Feebates, Rebates and Gas-Guzzler Taxes: A Study of Incentives for Increased Fuel Economy," *Energy Policy*, vol. 33, no. 6, pp. 721-827, 2004.

D.L. Greene and J. Hopson, "An Analysis of Alternative Forms of Automotive Fuel Economy Standards for the United States," *Transportation Research Record No. 1842*, pp. 20-28, Transportation Research Board, Washington, DC, 2003.

H.L. Hwang, S.M. Chin and D.L. Greene, "In, Out, Within and Through: Geography of Truck Freight in the Lower 48," *Transportation Research Record,* no. 1768, pp. 18–25, Transportation Research Board, Washington, DC, 2001.

D.L. Greene and S.E. Plotkin, "Energy Futures for the U.S. Transportation Sector," *Energy Policy*, vol. 29, no. 14, pp. 1255–1270, 2001.

D.L. Greene and N. Tishchishyna, "The Costs of Oil Dependence: A 2000 Update," *Transportation Quarterly*, vol. 55, no. 3, pp. 11–32, 2001.

H.L. Hwang, D.L. Greene, S.M. Chin, J. Hopson and A.A. Gibson, "Real-time Indicators of VKT and Congestion: One Year of Experience," *Transportation Research Record*, no. 1719, pp. 209–214, Transportation Research Board, Washington, DC, 2000.

D.L. Greene and J.M. DeCicco, "Engineering-Economic Analyses of Automotive Fuel Economy Potential in the United States," *Annual Review of Energy and the Environment*, vol. 25, pp. 477–536, 2000.

L.A. Greening, D.L. Greene and C. Difiglio, "Energy Efficiency and Consumption—The Rebound Effect—A Survey," *Energy Policy*, vol. 28, pp. 389–401, 2000.

R.N. Schock, W. Fulkerson, M.L. Brown, R.L. San Martin, D.L. Greene and J. Edmonds, "How Much Is Energy R&D Worth as Insurance?" *Annual Review of Energy and the Environment*, vol. 24, pp. 487–512, Annual Review, Palo Alto, California, 1999.

S.M. Chin, D.L. Greene, J. Hopson, H.L. Hwang and B. Thompson, "Towards Real-Time Indices of U.S. Vehicle Travel and Traffic Congestion," *Transportation Research Record*, no. 1660, pp. 132–139, National Academy Press, Washington, DC, 1999.

"Estimating the Fuel Economy Rebound Effect for Household Vehicles in the U.S.," *The Energy Journal*, vol. 20, no. 3, pp. 1–31, 1999.

"Survey Evidence on the Importance of Fuel Availability to Choice of Alternative Fuels and Vehicles," *Energy Studies Review*, vol. 8, no. 3, pp. 215–231, 1998.

"Why CAFE Worked," *Energy Policy*, vol. 26, no. 8, pp. 595–614, 1998.

D.L. Greene and Donald W. Jones and Paul N. Leiby, "The Outlook for U.S. Oil Dependence," *Energy Policy*, vol. 26, no. 1, pp. 55–69, 1998.

Erik F. Haites
Margaree Consultants Inc.
120 Adelaide Street West, Suite 2500
Toronto, Ontario, M5H 1T1
Tel: 416 369 0900; Fax: 416 369 0922

Education

1964	B.Sc. (Mathematics)	University of Alberta
1966	M.B.A.	McGill University
1969	M.S. (Economics)	Purdue University
1969	Ph.D. (Economics)	Purdue University

Research Interests

Market-based policies to reduce emissions and promote renewable energy, including emissions trading for conventional pollutants and greenhouse gases, Kyoto mechanisms, pollution taxes, renewable portfolio standards, and feed-in tariffs.

Professional Service/Activities

1995-present	President, Margaree Consultants
1989–1995	Principal, Barakat & Chamberlin
1983–1989	Vice President, The DPA Group Inc.
1981–1983	President, Middleton Associates
1977–1981	Self-Employed Consultant
1973–1977	Assistant Professor, University of Western Ontario
1972–1973	Senior Economist, Corporate Planning Department, Shell Canada
1969–1972	Senior Consultant, Stevenson and Kellogg Ltd.

Publications

Haites, E.F., F. Yamin, O. Blanchard, and C. Kemfert. 2004. Implementing the Kyoto Protocol without Russia. Climate Policy, vol. 4, no. 2, December, pp. 143-152.
Haites, E.F. 2004. Conclusion: Mechanisms, Linkages and the Direction of the Future Climate Regime. Part IV of The Kyoto Protocol Flexible Mechanisms: Implementation and Evolution within Europe and Worldwide, Farhana Yamin (ed.), Earthscan, London, pp. 327-356.
Haites, E.F. and F. Yamin. 2004. Overview of the Kyoto Mechanisms. International Review for Environmental Strategies, vol. 5, no. 1, pp. 199-215.

Haites, E.F. and F. Missfeldt. 2004. Liquidity Implications of a Commitment Period Reserve at National and Global Levels. Energy Economics, vol. 26, pp. 845-868.

Scott, M.J., J.A. Edmonds, N. Mahasenan, J.M. Roop, A.L. Brunello and E.F. Haites. 2004. International Emission Trading and the Cost of Greenhouse Gas Emissions Mitigation and Sequestration. Climatic Change, vol. 64, no. 3, June, pp. 257-287.

Kemfert, C., E.F. Haites and F. Missfeldt. 2003. Can Kyoto Protocol Parties Induce the United States to Adopt a More Stringent Greenhouse Gas Emissions Target? Interdisciplinary Environmental Review, vol. 5, no. 2, December, pp. 119-141.

Haites, E.F. 2003. Output-based Allocation as a Form of Protection for Internationally Competitive Industries. Climate Policy, vol. 3, supplement 2, December, pp. S29 - S41. Missfeldt, F. and E.F. Haites. 2002. Analysis of a Commitment Period Reserve at National and Global Levels. Climate Policy, vol. 2, no. 1, January, pp. 51-70.

Missfeldt, F. and E.F. Haites. 2001. The Potential Contribution of Sinks to Meeting Kyoto Protocol Commitments. Environmental Science and Policy, vol. 4, pp.269-292.

Haites, E.F. (Lead Author) with 20 other lead and contributing authors. 2001. Policies, Measures, and Instruments. Chapter 6 of Climate Change 2001: Mitigation, Contribution of Working Group III to the Third Assessment Report of the Intergovernmental Panel on Climate Change, Bert Metz, Ogunlade Davidson, Rob Swart and Jiahua Pan eds., Cambridge University Press, Cambridge.

Haites, E.F. and F. Missfeldt. 2001. Liability Rules for International Trading of Greenhouse Gas Emissions Quotas. Climate Policy, vol. 1, no. 1, January, pp. 85-108.

Haites, E.F. 2001. 'Bubbling' and the Kyoto Mechanisms. Climate Policy, vol. 1, no. 1, January, pp. 109-116.

Haites, E.F. and T. Hussain. 2000. The Changing Climate for Emissions Trading in Canada. RECEIL, Review of European Community & International Environmental Law, vol. 9, no. 3, pp. 264-275.

Haites, E.F. (Contributor) with 27 lead and 26 other contributing authors. 2000. Emissions Scenarios. Intergovernmental Panel on Climate Change, Cambridge University Press, Cambridge.

Haites, E.F. and F. Yamin. 2000. The Clean Development Mechanism: Proposals for its Operation and Governance. Global Environmental Change, vol. 10, no. 1, pp. 27-45.

Haites, E.F. and A. Proestos. 2000. Suitability of Non-Energy Greenhouse Gases for Emissions Trading. in J. van Ham, A.P.M. Baede, L.A. Meyer, and R. Yebma eds., Non-CO_2 Greenhouse Gases: Scientific Understanding, Control and Implementation, Kluwer Academic Publishers, Dordrecht, The Netherlands, pp. 417-424.

Bruce, J.P., M. Frome, H. Janzen, R. Lal, K. Paustian and E.F. Haites. 1999. Carbon Sequestration in Soils. Journal of Soil and Water Conservation, v. 54, n. 1, Jan.-Mar.

Hoffert, M.I. with E.F. Haites as one of 10 co-authors. 1998. Energy Implications of Future Stabilization of Atmospheric CO_2 Content. Nature, October 11.

Burke Hales
College of Oceanic and Atmospheric Sciences
Oregon State University
Corvallis, Oregon

Education

1995 Ph.D. University of Washington, School of Oceanography
1992 M.S. University of Washington, School of Oceanography
1988 B.S. University of Washington, College of Engineering

Research Interests

Coastal Oceanography: Analysis and synthesis of the physics, biology, and chemistry of the coastal ocean, using observations collected with high-speed sampling and analysis systems.

Mesoscale Surface Ocean Processes: Analysis and synthesis of the physics, biology, and chemistry of the surface ocean, using observations collected with high-speed sampling and analysis systems.

Analytical Environmental Chemistry: Development of sensors and systems for high-speed and robust measurement of ocean chemistry.

Benthic Biogeochemistry: *In situ* field measurements of sediment pore water chemistry and numerical models of transport and chemical kinetics in sediments.

Employment History

2004-Present Associate Professor, College of Oceanographic and Atmospheric Sciences, Oregon State University

1998-2004 Assistant Professor, College of Oceanographic and Atmospheric Sciences, Oregon State University

1998-present Adjunct Associate Research Scientist, Lamont-Doherty Earth Observatory of Columbia University

1997-1998 Associate Research Scientist, Lamont-Doherty Earth Observatory of Columbia University

1995-1997 Postdoctoral Research Fellow, Lamont-Doherty Earth Observatory of Columbia University (Postdoctoral Advisor: Dr. Taro Takahashi)

Publications

Vaillancourt, R., J. Marra, R. Houghton, L. Prieto, B. Hales, and D. Hebert, 2005. Light absorption by particles and CDOM at the New England shelfbreak front during Summer. *G-Cubed*, submitted.

Hales, B., L. Karp-Boss, A. Perlin, and P. Wheeler, 2005. Oxygen production and carbon sequestration in an upwelling coastal margin, *Global Biogeochemical Cycles*, submitted

Hales, B., J. N. Moum, P. Covert, and A. Perlin, 2005. Irreversible Nitrate Fluxes Due To Turbulent Mixing in a Coastal Upwelling System. *Journal of Geophysical Research—Oceans*, in press.

Bandstra, L., B. Hales, and T. Takahashi, 2005: High-frequency measurement of seawater total carbon dioxide. Submitted to *Mar. Chem.*

Chase, Z., B. Hales, and T. Cowles, 2005. Distribution and variability of iron input to Oregon coastal waters during the upwelling season *Journal of Geophysical Research—Oceans*, In press

Hales, B., T. Takahashi and L. Bandstra, 2005. Atmospheric CO_2 uptake by a coastal upwelling system *Global Biogeochem. Cycles* 19, doi:10.1029/2004GB002295

Hales, B., D. Chipman and T. Takahashi, 2005. High-frequency measurement of partial pressure and total concentration of carbon dioxide in seawater using microporous hydrophobic membrane contactors. *Limnology and Oceanography: Methods* **2**, 356-364.

Karp, L., P. Wheeler, B. Hales, and P. Covert, 2004. Distributions and variability of POM in a coastal upwelling system. *Journal of Geophysical Research—Oceans* **109**, C09010, doi:10.1029/2003JC002184.

Hales, B., and T. Takahashi, 2004. High-resolution biogeochemical investigation of the Ross Sea, Antarctica, during the AESOPS (U. S. JGOFS) program. *Global Biogeochemical Cycles* **18**, GB3006, doi:10.1029/2003GB002165.

Coale, K. H., and others; (Hales is 13th in list of 42 co-authors), 2003. Southern Ocean Iron Enrichment Experiment (SOFeX): Iron, Silicon and Light Interactions in Antarctic Waters. *Science* **304,** 408-414

Hales, B, Takahashi, and van Geen, A. 2004. High-frequency measurement of seawater chemistry: Flow-injection analysis of macronutrients. *Limnology and Oceanography: Methods* **2,** 91–101

Hales, B., 2003. Respiration, dissolution, and the lysocline. *Paleoceanography*, **18**, 1099-1113

Hales, B. and T. Takahashi, 2002. The Pumping Seasoar: A high resolution seawater sampling platform. *J. Tech.* **19**, 1096-1104

Hales, B., Sweeney , C., and Takahashi, T., 2001. Small-scale variability in the Ross Sea. *Oceanography* **14**, 90-91.

Alleau Y., Colbert D., Covert P., Haley B., Qiu X., Collier R., Falkner K., Hales B., Prahl, F. and Gordon L., 2001. Th-234 applied to particle removal rates from the surface ocean: a mathematical treatment revisited. *Geophys. Res. Letters* **28**, 2855-2857.

Elisabeth Huber-Sannwald
Department of Environmental Engineering and Natural Resource Management
Instituto Potosino de Investigación Científica y Tecnológica
San Luis Potosí, S.L.P. 78216, Mexico
Tel: +52 (444) 834 2000

Education

1996	Ph.D.	Range Ecology, Utah State University, Logan, Utah,
1990	M.S.	Biological Sciences, Innsbruck University, Innsbruck, Austria

Employment History

2001- present   Associate Professor, Department of Environmental Engineering and Natural Resource Management, IPICYT, San Luis Potosí, México

1998-2001   Research Assistant, Department of Grassland Science, Center of Life Sciences, Technical University Munich, Weihenstephan, Germany

1997-1998   Scientific Officer, Focus 4 of international GCTE core project of IGBP, Institute of Ecology, University of Buenos Aires, Buenos Aires, Argentina

1996-1997   Research Assistant, Utah State University, Logan, Utah, USA

1991-1996   Research Graduate Assistant, Utah State University, Logan, Utah, USA

1991-1992   Research Assistant, Utah State University, Logan, Utah, USA

1986-1991   Research Assistant, Coordination of Environmental Assessment and Inventory of endangered ecosystems, University of Innsbruck, Innsbruck, Austria

Professional Service/Activities

2002   Sistema Nacional de Investigadores Nivel 1, Mexico

2001   Cátedra Patrimonial Nivel II, Mexico

1997   Don Dwyer, Award for Scientific Excellence as Graduate Student (PhD), Rangeland Resources Department, Utah State University

1991   Fulbright Scholarship, Austria

1987-1989 Endowment Scholarship of University of Innsbruck, Austria

Publications and Presentations

Huber-Sannwald, E. and D.A. Pyke. (2005) *In press*. Establishing native grasses in a big sagebrush dominated site: An intermediate restoration step. Restoration Ecology.

Herben, T. and E. Huber-Sannwald. 2002. Effect of management on species richness of grasslands: sward-scale processes lead to large-scale patterns. Grassland Science in Europe, Vol. 7: 625-643.

Chapin, III. F.S., O.E. Sala, E. Huber-Sannwald (eds). 2001. Global Biodiversity in a Changing Environment. Springer-Verlag, New York, p. 376.

Chapin, III, F.S., O.E. Sala, E. Huber-Sannwald, and Rik Leemans (2001). The future of biodiversity in a changing world. In: F.S. Chapin, III., O.E. Sala, E. Huber-Sannwald (eds). Global Biodiversity in a Changing Environment. Springer-Verlag, New York, pp. 1-4.

Sala, O.E., E. Huber-Sannwald, and F.S. Chapin, III (2001). Conclusions. In: F.S. Chapin, III., O.E. Sala, E. Huber-Sannwald (eds). Future Scenarios of Global Biodiversity. Springer-Verlag, New York, pp. 351-367.

Huber-Sannwald, E. and R.B. Jackson. (2001). Heterogenous soil resource distribution and plant responses - from individual-plant growth to ecosystem functioning. In: K. Esser, U. Lüttge, J.W. Kadercit, W. Beyschlag (eds) Progress in Botany 62: 450-475.

Sala, O.E., F. S. Chapin III, J.J. Armesto, E. Berlow, J. Bloomfield, R. Dirzo, E. Huber-Sannwald, L.F. Huenneke, R. B. Jackson, A. Kinzig, R. Leemans, D. M. Lodge, H. A. Mooney, M. Oesterheld, N. LeRoy Poff, M. T. Sykes, B. H. Walker, M.Walker, D. H. Wall. (2000). Global Biodiversity Scenarios for the year 2100, Science 287: 1770-1774.

Huber-Sannwald, E., D. A. Pyke, M.M. Caldwell and S. Durham (1999). Interactions between clonal grasses under heterogeneous environmental conditions in a semiarid desert of the Great Basin, USA. Bielefelder Ökologische Beiträge, 14, pp. 41-51.

Huber-Sannwald, E., D. A. Pyke, and M. M. Caldwell. 1998. Clonal plasticity of a rhizomatous grass in heterogeneous environments: influence of nutrient patches and neighboring plant root systems. Ecology 79 (7): 2267-2280.

Huber-Sannwald, E., D. A. Pyke, and M. M. Caldwell. (1997). Perception of neighbouring root systems by rhizomes and roots: morphological manifestations of interacting clonal plants. Canadian Journal of Botany 75: 2146-2157.

II. Bi-national Ecological Meeting, Mendoza, Argentina, "A new paradigm of desertification combining the biophysical and socioeconomic dimension: Biophysical aspects at different spatial scales; 2004.

I. ARIDnet Workshop in Latinamerica: Land degradation in semiarid regions of the Americas. The Amapola, Mexico Case Study. IPICYT, San Luis Potosi, Mexico. Titulo: Desertification: When, why and how does it occur? 2004

1st National Week of Supercomputers, San Luis Potosí, Mexico; "The Inter-American Atmosphere/Biosphere (IANABIS) Project", 2004.

XIV Annual Symposium Jornada; University of New Mexico, NM, Las Cruces, July, 2004; "GRACILIS" - The needs and objectives of a research network for semiarid grasslands in Mexico.

MARK K. JACCARD
School of Resource and Environmental Management
Simon Fraser University
Vancouver, B.C., CANADA, V5A 1S6
Tel. (604) 291-4219

EDUCATION:

Ph.D.: University of Grenoble, Department of Economics / Institute of
    Energy Economics and Policy, 1987.
Masters of Natural Resources Management: Simon Fraser University, 1984.
Bachelor of Arts: Simon Fraser University, 1978.

PROFESSIONAL EXPERIENCE:

1986 - 2003 to present:
SIMON FRASER UNIVERSITY
Full Professor, School of Resource and Environmental Management. Teaching graduate courses in ecological economics, environmental policy, energy management and policy, and energy system modelling. Supervision of graduate student research. Director of Energy and Materials Research Group

1992-1997:
BRITISH COLUMBIA UTILITIES COMMISSION
Chairman and Chief Executive Officer. Director of a quasi-judicial regulatory body charged with regulating the rates and investments of all energy utilities in B.C. Responsibilities split between administration of the commission and role as chairperson for public hearings and regulatory decisions. On leave from university for administration and teaching duties, but sustained research program and student thesis supervision.

RECENT PUBLICATIONS – 2002 - 2005

Jaccard, M. Sustainable Fossil Fuels: The Unusual Suspect in the Quest for Clean and Enduring Energy (forthcoming – Cambridge University Press).

Washbrook, K., Haider, W. and M. Jaccard, "Estimating Commuter Mode Choice: a Discrete Choice Analysis of the Impact of Road Pricing and Parking Charges," (forthcoming – Transportation).

Nyboer, J. and M. Jaccard, "Simulating Policies to Induce Technological Change: The Usefulness of Energy-Economy Models Under Technological and Behavioral Uncertainty," (forthcoming – International Journal of Energy Technology and Policy).

Jaccard, M. and M. Dennis, "Estimating Home Energy Decision Parameters for a Hybrid Energy-Economy Policy Model," (forthcoming - Environmental Modeling and Assessment).

Jaccard, M. "Policies that Mobilize Producers Toward Sustainability: The Renewable Portfolio Standard and the Vehicle Emission Standard," In: Building Canadian Capacity:

<u>Sustainable Production and the Knowledge Economy</u>, G. Toner (ed.), (forthcoming – UBC Press).

Rivers, N. and M. Jaccard, "Canada's Efforts Towards Greenhouse Gas Emission Reduction: A Case Study on the Limits of Voluntary Action and Subsidies," <u>International Journal of Global Energy Issues</u>, V.23, 4, 2005, 307-323.

Rivers N. and M. Jaccard, "Useful Models for Simulating Policies to Induce Technological Change," (forthcoming – <u>Energy Policy</u>).

Jaccard, M. "Hybrid Energy-Economy Models and Endogenous Technological Change," In R. Loulou, J-P Waaub and G. Zaccour (eds.) <u>Energy and Environment,</u> New York: Springer, 2005, 81-110.

Horne, M., Jaccard, M. and K. Tiedemann, "Improving Behavioral Realism in Hybrid Energy-Economy Models Using Discrete Choice Studies of Personal Transportation Decisions," <u>Energy Economics</u>, V27, 2005, 59-77.

Rivers, N. and M. Jaccard, "Combining Top-down and Bottom-up Approaches to Energy-economy Modeling Using Discrete Choice Methods," <u>The Energy Journal</u>, V26, N.1, 2005, 83-106.

Jaccard, M., Murphy, R. and N. Rivers, "Energy-Environment Policy Modeling of Endogenous Technological Change with Personal Vehicles: Combining Top-Down and Bottom-Up Methods," <u>Ecological Economics</u>, V51, 2004, 31-46.

Jaccard, M., Rivers, N. and M. Horne, <u>The Morning After: Optimal GHG Policies for Canada's Kyoto Obligation and Beyond</u>, Toronto: CD Howe Institute, 2004, 31 pages.

Jaccard, M. "Greenhouse Gas Abatement: Controversies in Cost Assessment," In C. Cleveland (ed.) <u>Encyclopedia of Energy</u>, New York: Elsevier, V.3, 2004, 57-65.

Jaccard, M. "Renewable Portfolio Standard," In C. Cleveland (ed.) <u>Encyclopedia of Energy</u>, New York: Elsevier, V.5, 2004, 413-421.

Jaccard, M. and C. Bataille, "If Sustainability is Expensive, What Roles for Business and Government? A Case Study of Greenhouse Gas Reduction Policy in Canada," <u>Journal of Business Administration and Policy Analysis,</u> V30-31, 2004, 149-183.

Weidou, N., Johansson, T., Wang, J., Wu, Z., Mao, Y., Zhu, Q., Zhou, F., Li, Z., Anderson, B., Farinelli, U., Jaccard, M. and R. Williams, "Transforming coal for sustainability: a strategy for China," <u>Energy for Sustainable Development</u>, V.VII, N.4, 2003, 21-30.

Murphy, R. and M. Jaccard, "The Voluntary Approach to Greenhouse Gas Reduction: A Case Study of BC Hydro," <u>Energy Studies Review</u>, V11, N2, 2003, 131-151.

Rivers, N., Jaccard, M., Tiedemann, K. and J. Nyboer, "Confronting the Challenge of Hybrid Modeling: Using Discrete Choice Models to Inform the Behavioural Parameters of a Hybrid Model," <u>American Council for an Energy-Efficient Economy: ACEEE 2003</u>, V1, 2003, 181-192.

Jaccard, M., Nyboer, J., Bataille, C. and B. Sadownik, "Modeling the Cost of Climate Policy: Distinguishing Between Alternative Cost Definitions and Long-Run Cost Dynamics," <u>The Energy Journal</u>, V24, N1, 2003, 49-73.

Jaccard, M., Loulou, R., Kanudia, A., Nyboer, J., Bailie, A. and M. Labriet, "Methodological Contrasts in Costing GHG Abatement Policies: Optimization and Simulation Modeling of Micro-Economic Effects in Canada," European Journal of Operations Research, V145,N1, 2003, 148-164.

Sadownik, B. and M. Jaccard, "Shaping Sustainable Energy Use in Chinese Cities," DISP 151, V.4, 2002, 15-22.

Jaccard, M., "Energy Planning and Management: Methodologies and Tools," in Encyclopedia of Life Support Systems, Oxford, UK: UNESCO, EOLSS Publishers, 2002.

Jaccard, M., Nyboer, J. and B. Sadownik, The Cost of Climate Policy, Vancouver: UBC Press, 2002, 242 pages.

Jaccard, M. and Y. Mao, "Making Markets Work Better," in Johansson and Goldemberg (eds.) Energy for Sustainable Development: A Policy Agenda, New York: United Nations Development Program, 2002, 41-77.

Nanduri, M., Nyboer, J. and M. Jaccard, "Aggregating Physical Intensity Indicators: Results of Applying the Composite Indicator Approach to the Canadian Industrial Sector," Energy Policy, V30, 2002, 151-163.

Jaccard, M., California Shorts a Circuit: Should Canadians Trust the Wiring Diagram? Toronto: C.D. Howe Institute, 2002, 28 pages.

Jennifer C. Jenkins
Gund Institute for Ecological Economics
University of Vermont
School of Natural Resources
590 Main Street
Burlington, VT 05405
Tel: (802) 656-2953; Fax: (802) 656-2995

Education

1991   B.A.   Biology, Dartmouth College
1995   M.F.S.   Forest Science, Yale University
1998   Ph.D.   Ecosystem Ecology, University of New Hampshire

Employment History

2002-present   Visiting Assistant Professor, Gund Institute for Ecological Economics, University of Vermont Rubenstein School of Environment and Natural Resources, Burlington, VT.
1998-2002   Research Forester, USDA Forest Service Northeastern Research Station Northern Global Change Program and Forest Inventory and Analysis

Professional Service/Activities

Delegate - National Academy of Sciences Workshop on Direct and Indirect Human Contributions to Terrestrial Greenhouse Gas Fluxes
U.S. Technical Expert - IPCC Working Group on Methodologies to Factor Out Direct Human-Induced Changes in Carbon Stocks and Greenhouse Gas Emissions by Sources and Removal by Sinks
Member - NCEAS Working Groups: Carbon Balance of North America and Eurasia; Development of a Consistent Global NPP database
Participant - Cary Conference IX: Understanding Ecosystems: The Role of Quantitative Models in Observation, Synthesis, and Prediction
Journal reviews - Canadian Journal of Forest Research, Climatic Change, Computers in Science and Agriculture, Ecological Applications, Ecosystems, Environmental Pollution, Forest Science, Global Change Biology, Journal of Biogeography, Mitigation and Adaptation Strategies for Global Change
Grant reviews - EPA STAR Fellowship Panel, NSF Long-term Research in Environmental Biology (LTREB) (2002), NSF Ecosystems, NASA New Investigator Program

Publications

Jenkins, J.C., D.C. Chojnacky, L.S. Heath and R.A. Birdsey. 2003. National-scale biomass estimators for United States tree species. Forest Science 49(1):12-35.

Jenkins, J.C., D.C. Chojnacky, L.S. Heath and R.A. Birdsey. 2003. A comprehensive database of biomass equations for North American tree species. USDA Forest Service General Technical Report NE- XXX (in review).

Pan, Y., J. Hom, J.C. Jenkins and R.A. Birdsey. 2003. Importance of foliar nitrogen concentration to predict forest productivity spatially across the Mid-Atlantic region. Forest Science (in press).

Smith, J, L.S. Heath and J.C. Jenkins. 2003. Forest volume-to-biomass models and estimates of mass for live and standing dead trees of US forests. Newtown Square, PA, USDA Forest Service General Technical Report NE-298. 57 p.

Jenkins, J.C. and R. Riemann. 2003. What does nonforest land contribute to the global C balance? Proceedings, Third Annual FIA Science Symposium, Traverse City, MI, Oct. 14-16, 2001 (in press).

Goodale, C.L., M.J. Apps, R.A. Birdsey, C.B. Field, L.S. Heath, R.A. Houghton, J.C. Jenkins, G.H. Kohlmaier, W. Kurz, S. Liu, G-J Nabuurs, S. Nillson and A. Shvidenko. 2002. Forest carbon sinks in the northern hemisphere. Ecological Applications 12:891-899.

Jenkins, J.C., R.A. Birdsey and Y. Pan. 2001. Biomass and NPP estimation for the mid-Atlantic (USA) region using plot-level forest inventory data. Ecological Applications 11:1174-1193.

Caspersen, J.P., S.W. Pacala, J.C. Jenkins, G.C. Hurtt, P.R. Moorcroft and R.A. Birdsey. 2000. Carbon accumulation in eastern U.S. forests is caused overwhelmingly by changes in land use rather than CO_2 or N fertilization or climate change. Science 290:1148-1151.

Hicke, J.A., G.P. Asner, J. Randerson, S. Los, R.A. Birdsey, J.C. Jenkins, C. Tucker and C. Field. 2002. Trends in North American net primary productivity derived from satellite observations, 1982-1998. Global Biogeochemical Cycles 16(2): 0.1029/2001GB001550.

Nemani, R.R., M.A. White, K. Nishida, S. Reddy, J.C. Jenkins and S.W. Running. 2002. Recent trends in hydrologic balance have enhanced the terrestrial carbon sink in the United States. Geophysical Research Letters 2002GL014867.

Jenkins, J.C., D.W. Kicklighter and J.D. Aber. 2000. Predicting the regional impacts of increased CO_2 and climate change on forest productivity. Pp. 383-423 In Responses of Northern U.S. Forests to Environmental Change, R.A. Birdsey, R.H. Mickler and J. Hom (eds). Springer-Verlag, New York.

Jenkins, J.C., D.W. Kicklighter, S.V. Ollinger, J.D. Aber, J.D. and J.M. Melillo. 1999. Sources of variability at a regional scale: A comparison using PnET-II and TEM 4.0 in northeastern U.S. forests. Ecosystems 2:555-570.

Mark H. Johnston
Saskatchewan Research Council
15 Innovation Blvd, Saskatoon, SK Canada S7N 2X8
Tel: (306) 933-8175; Fax: (306) 933-7817

Education

1976	B.S.	Forest Resources Management, University of Minnesota
1981	M.S.	Forest Science (Fire Ecology), University of Alberta
1990	Ph.D.	Tropical Forest Ecology, State University of New York

Research interests

Integration of carbon sequestration options with forest management; opportunities for afforestation in agricultural landscapes; modeling impacts of climate change on forest ecosystems and forest management; role of disturbance in forest ecosystems and options for disturbance emulation in forest management; use of fast-growing woody species for bioenergy and biofuels.

Employment History

1992-1997	Research Scientist, Ontario Ministry of Natural Resources
1997-2001	Manager, Forest Science Programs, Saskatchewan Environment
2001-present	Senior Research Scientist, Saskatchewan Research Council

Publications

Johnston, M. and Williamson, T. 200X. Climate change and its implications for forest stand yields and soil expectation values: a northern Saskatchewan case study. Invited paper, special issue of Forestry Chronicle on climate change and forest management (in prep.).

Drew, A.P., Zhao, Y., Johnston, M.H. and Weaver, P.L. 200X. Fifty-five years of change in rain forest structure and composition at El Verde, Puerto Rico. Journal of Tropical Ecology (Submitted).

Wilson, S.J., Carleton, T.J., Johnston, M.H., and Elliott, J.A. 200X. Response of a boreal mixedwood community to experimental fire, clear-cutting, clearcutting followed by fire, and simulated blowdown followed by fire. Forest Ecology and Management (in press).

Kulshreshtha, S., Johnston, M. and Lac, S. 2003. Value of stored carbon in protected areas: A case study of Saskatchewan Provincial Parks. Prairie Forum 28:127-144.

Lemprière, T., Johnston, M., Willcocks, A., Bogdanski, B., Bisson, D., Apps, M., Bussler, O. 2002. Saskatchewan forest carbon sequestration project. Forestry Chronicle 78:843-849.

Dore, M., Kulshreshtha, S., and Johnston, M. 2001. An integrated economic – ecological analysis of land use decisions in forest-agriculture fringe region of northern Saskatchewan. Geographical and Environmental Modelling 5:159-175.

Johnston, M., Kulshreshtha, S., and Baumgartner, T. 2001. Agroforestry in the prairie landscape: opportunities for climate change mitigation through carbon sequestration. Prairie Forum 25:195-213.

Dore, M. and Johnston, M. 2001. Value of carbon storage in Canadian forests. Journal of Sustainable Forestry. 12:123-151.

Johnston, M. and Uhlig, P. 2000. Carbon storage in soils and vegetation among forested ecosystem types in northern Ontario. Pp. 63-74 in: Sustainable Forest Management And Global Climate Change: Selected Case Studies from the Americas, Dore, M.H. and Guevara, R., Eds. Edward Elgar Publishing, Cheltenham, UK.

Dore, M. and Johnston, M. 2000. The carbon cycle and the value of forests as a carbon sink: a boreal case study. Pp. 79-106 in: Sustainable Forest Management And Global Climate Change: Selected Case Studies from the Americas, Dore, M.H. and Guevara, R., Eds. Edward Elgar Publishing, Cheltenham, UK.

Johnston, M. and Elliott, J. 1998. The effect of fire severity on ash, plant and soil nutrient levels following experimental burning in a boreal mixedwood stand. Canadian Journal of Soil Science 78:35-44.

Patricia Romero Lankao
Department of Politics and Culture, Autonomous Metropolitan University (UAM), Xochimilco,
Calzada del Hueso 1100, Villa Quietud, 04960, Mexico City
Tel: (525) 4837110; Fax: (525) 5949100

Education

1998   Ph.D.   University of Bonn, Germany, Agricultural Sciences and Environmental
        Policy
1997   Ph.D.   Autonomous Metropolitan University (UAM-X), Regional Studies   1991   M.
Sc.   UNAM, Sociology
1986   B.S.   UNAM, Sociology

Research Interests

Participation in several international committees designing a research agenda for the human
dimensions of global change such as cities, water and industrial transformation (International
Human Dimensions Program on Global Environmental Change, IHDP).

Since 2002 member of the Steering Scientific Committee of the Global Carbon Project
sponsored by IHDP, the International Geosphere-Biosphere Program (IGBP) and the World
Climate Research Program (WCRP), and in charge of designing and implementing a research
agenda.

 Teaching in Neoliberalism and the Environment (grad. seminar, LA_S 595D with Diana Liverman),
Environmental Economy and Policy (grad. seminar, Postgraduate Studies in Economy, UAM),
History of Mexico (gen. ed. UAM-Xochimilco), Sociology of Science (gen. ed. UAM-Xochimilco),
Classics of Social Theory (grad. seminar, UAM-Xochimilco), Rural development and the
environment (grad. seminar, UAM-Xochimilco), urban development and the environment (grad.
seminar, UAM-Xochimilco), Environmental Policy in Mexico (grad. seminar, UNAM).

Participation in the Work Group II Contribution to the IPCC Fourth Assessment Report in two
tasks: Co-Lead Author of the 7th Chapter "Industry, Settlement and Society", member of the tem
in charge of the IPCC Working Group II and III on "the cross-cutting themes Adaptation-
Mitigation and Sustainable Development"

Publications

*Romero, P., "Impacto Socioambiental, en Xochimilco y Lerma, de las obras de abastecimiento
de la ciudad de México", (Social and environmental impact, in Xochimilco and Lerma, of
Mexico City's water supply systems), México, UAM- Xochimilco, 1993, (National Environmental
Prize in 1992).*

*Romero, P., "Probleme der Realisierung umweltpolitischer Strategien in der mexikanischen
Landwirtschaft" (Implementation problems of environmental policy strategies in Mexican
Agriculture), Bonn Germany, Wehle Witterschlick, 1998.*

Romero, P. "Obra hidráulica en la ciudad de México y su impacto socioambiental, 1880-1990", (Water system in Mexico City, Social and Environmental Impact. 1880-1990). México, Instituto Mora, 1999.

Romero, Duffing, López, "Mexico: Who is Doing What in Human Dimensions of Global Environmental Change?, IHDP-UAM, Germany/Mexico, bilingual document, 2001.

Romero, P. "Política Ambiental Mexicana. Distancia entre objetivos y logros" (Mexican Environmental Policy. Distance between goals and results), México, UAM-Xochimilco, 2001. Romero, P., "Abordaje y simplificación neoclásica de lo ambiental" (Neo-classical analysis and simplification of environmental issues), in: Delgado, J.: et.al. (Coord.) *La formación territorial de la ciudad de México*, México, 1999, UAM-X/Plaza y Valdes.

Romero, P., "Visión Gubernamental del Cambio Ambiental, los Desastres y la Vulnerabilidad" (Governmental vision of environmental change, disasters and vulnerability), in: La Nación ante los Desastres. Retos y Oportunidades Hacia el Siglo XXI, México, SG, 1999, 119-128.

- Varady, R. Romero, P. y Hankins, K. "Whither Hazardous-materials Management in the U.S.-Mexico Border Region?", in Both Sides of the Border. Transboundary Environmental Management Issues Facing Mexico and United States *edited by Fernandez, L. y Carson, R.T. Dordrecht, Kluwer Academic Publishers, 2002.*

Romero, P. "Agua en el Alto Lerma. Experiencias y lecciones de uso y gestión" (Water in Lerma. Experiences and Lessons of Use and Management), in I Encuentro de Investigadores de la Cuenca Lerma Chapala Santiago, *México, Colmich-U. of Guadalajara.*

Romero Lankao, P. 2003. "Pathways of regional development and the carbon cycle" in *Toward CO2 Stabilization: Issues, Strategies, and Consequences*, edited by Field, C. and Raupach M. Island Press (in press).

Raupach, M. and 11 others 2003. "Atmospheric Stabilization in the Context of Carbon-climate-Human Interactions" in *Toward CO2 Stabilization: Issues, Strategies, and Consequences*, edited by Field, C. and Raupach M. Island Press (in press).

Romero, P., "Evaluación de impacto ambiental: instrumento de política pública", (Environmental Impact Assessment: a public policy instrument in Mexico), in: *Revista Argumentos* 21, 1994, 7-21.

Romero, P., "Sustainability and Public Management Reform: Two Challenges for Mexican Environmental Policy", in American Review of Public Administration, 2000. - Varady, R. Romero, P. y Hankins, K. 2001, "Managing Hazardous Materials along the US-Mexico Border", in *Environment*, Vol.43, No. 10, pp.22-36.

Wilder, M. and Romero Lankao, P. (2004), "Paradoxes of Decentralization: Water Reform in and its Social Implications in Mexico", submitted to World Development.

Professional Service/Activities

- Scholarship, Mexican Council for Science and Technology (CONACYT), first Ph.D., Regional Studies, 1992-1994
- National Environmental Prize (Premio Serfín del Medio Ambiente), 1992
- Member of the Mexican National System of Researchers: during 1994-1998 as Candidate; during 1998-2000 level I, and since 2001, level II
- Scholarship, Konrad-Adenauer-Stiftung, second Ph.D. in Agricultural Sciences and Environmental Policy at Bonn University, Germany, 1995-1998. Complementary scholarship, Ford, John D. and Catherine T. MacArthur, and William and Flora Hewlett Foundations, field survey in Mexico for 2nd Ph.D. in Agricultural Sciences and Environmental Policy, 1996
- Fellow, 8th cohort of the Program Leadership for Environment and Development (LEAD International), since 1999.
- Grant from AIACC to undertake, as Co-Pi, research proposal on "Integrated Assessment of Social Vulnerability and Adaptation to Climate Variability and Change among Farmers in Mexico and Argentina" 2003.
- Research grant from START and NCAR to undertake the project "How Local Governments Manage Global Warming? Institutional Settings and Carbon in Mexico-City" (2003-2004)
- Research grant from IAI to organize two workshops aimed at undertaking the project Can Cities Reduce Global Warming? Urban development and the carbon cycle in Latin America" (2004-2005)
-Grant from the National Institute of Ecology to work on the project "Towards a Program to Foster Research on Climate Change in Mexico", (2004)
- Nomination to write the chapter "Human Settlements, Energy and Industry" within the IPCC Working Group II Fourth Assessment Report (2004-2007).

JAMES E. MCMAHON
Head, Energy Analysis Department
Lawrence Berkeley National Laboratory
One Cyclotron Road, M/S 90-4000
Berkeley, CA 94720
Tel. (510) 486-6049

EMPLOYMENT HISTORY
1978-present: Lawrence Berkeley National Laboratory, Berkeley, California
Head, Energy Analysis Department
Leader, Energy Efficiency Standards Group
Staff Scientist, Energy Analysis Program, Environmental Energy Technologies
    Division
1975-1978: Postdoctoral fellow, Department of Chemistry, University of
    California, Berkeley

SELECTED CONSULTING EXPERIENCE

ADEME, Paris (France)
Australia Greenhouse Office (Australia)
Centre universitaire d'etude des problemes de l'energie, Geneve, (Switzerland)
Danish Energy Agency, Copenhagen (Denmark)
ENEA, Rome (Italy)
IEA, Paris (France)
Barakat & Chamberlin, Inc, Oakland, California
Morse, Richard, & Weisenmiller & Associates, Oakland, California,
Regional Economic Research, Inc., San Diego, California
United Nations, New York
Xenergy, Oakland, California

PROFESSIONAL SOCIETIES

American Association for the Advancement of Science

RELEVANT RESEARCH EXPERIENCE

Manage Energy Analysis Department. Provide scientific and strategic leadership for the world's largest and most experienced group of energy analysts, whose mission is to compile and analyze energy information for the residential, commercial, industrial, and transportation sectors. Represent LBNL in eleven-laboratory Energy Water Nexus, exploring interactions between energy and water systems.

Manage technical and economic analysis of US efficiency standards for appliances, lighting and equipment. Modify, debug, and maintain detailed databases and computer models of U.S. energy consumption in residential and commercial buildings for U.S. Department of Energy studies related to appliance, lighting and equipment standards. Provide technical support documents containing full documentation of analyses.

Consult for energy companies and non-US governments regarding energy efficiency policies. Advise government agencies (including Australia, China, European Union, Ghana, India). Transfer databases and energy demand forecasting models to researchers within and outside US. Review load forecasts and demand reduction program plans for utilities and their consultants. Include impacts of US appliance energy efficiency standards and utility incentive programs.

Review energy surveys and forecasting models. Review Buildings Module for US Department of Energy National Energy Modeling System. Review Energy Information Administration RECS and CBECS surveys. Provide design suggestions, reviews and default data inputs for energy demand forecasting models, such as REEPS (residential) and COMMEND (service/commercial).

Analyzed U.S. Department of Energy (DOE) Efficiency Standards for Appliances, Lighting and Equipment. Managed economic and technical analysis of proposed and adopted national standards, including engineering analysis and impacts on consumers, manufacturers, utilities, and environment (analyst since 1979, manager since 1986). Designed, maintained, and executed residential and commercial sector energy demand forecasting models. Performed studies of technological feasibility and economic justification for DOE efficiency standards for 12 residential appliance types (including furnaces, boilers, water heaters, air conditioners, and all major appliances); lighting equipment (including fluorescent lamp ballasts and high intensity discharge lamps); commercial heating, cooling and water heating equipment; distribution transformers and small electric motors. Provided estimates of energy savings, changes in purchase and usage patterns, and net economic benefit to residential and commercial consumers due to proposed policy. Invited participant to DOE conference "Estimating the Benefits of Government-Sponsored Energy R&D," March, 2002. Co-authored with S. Wiel, "Energy-Efficiency Labels and Standards: A guidebook for appliances, equipment and lighting," 2005.

Analyzed Other National Energy Policies. Managed analyses of policy options (including labels, information and rating programs, and/or performance) regarding office equipment, lamps, luminaires, small electric motors, and plumbing products. Coauthor of major studies, including Evaluation of Advanced Technologies, Early Replacement of Appliances, Potential for Electricity Efficiency Improvements in the US Residential Sector, Federal Policy Options for Lighting Efficiency. Contributor to National Commission on Energy Policy report, "Ending the Energy Stalemate," 2004.

Behavior of the Market for Efficient Appliances. Analyzed the effects of federal, state and utility programs on efficiency of new appliances sold nationally and regionally. Contributed to analyses of market behavior regarding appliance efficiency choice, for all major residential appliances.

Analysis of Commercial Building Energy Consumption. Analyzed full set of policy options for improving indoor lighting energy consumption, including education/information programs (for consumers and designers, and component labeling), national incentive programs (rebates and tax credits), voluntary component standards, mandatory system performance standards (i.e., building codes), and mandatory component performance standards. Reviewed published works on energy consumption in commercial buildings; analyzed energy consumption methodologies; coauthored report with national energy policy implications.

Energy Demand Models. Reviewer for USDOE/EIA National Energy Modeling System (NEMS). Collaborated closely with EPRI on design of, and data required for, residential and commercial end-use energy forecasting models, REEPS and COMMEND. Improved LBNL's national residential model, including: applied vintage structure and historical shipments data to turnover of appliances; incorporated recent improvements in appliance efficiencies and housing construction techniques; expanded coverage of end uses; replaced decision algorithm for appliance efficiency choice; improved representation of equipment cost/price data; converted from average value to distribution of appliance efficiencies in policy cases; and updated national data base.

Ronald B. Mitchell
Department of Political Science
University of Oregon
Eugene OR 97403-1284
Tel: (541) 346-4880; Fax: (541) 346-4860

Education

1992   Ph.D.   Harvard University, Graduate School of Arts and Sciences, Public Policy 1985
M.P.P. Harvard University, John F. Kennedy School of Govt., Public Policy
1981   B.A.   Stanford University, American Studies

Employment History

1993 – present   Assistant and Associate Professor, University of Oregon, Department of
Political Science
1999-2001   Visiting Associate Professor, Stanford University, Center for
Environmental Science and Policy and International Policy Studies
Program

Professional Service/Activities

Member – State of Oregon Governor's Advisory Group on Global Warming
Member – National Research Council, Committee on the Human Dimensions of
Global Change
Chair of Working Group 5 on Socio-Economic Aspects, Climate Impacts on Oceanic TOp
Predators (CLIOTOP) Project
Member – DIVERSITAS – Scientific Committee for Core Project on
"bioSUSTAINABILITY: Conservation and Sustainable Use of Biodiversity,"
International Council for Science
Editorial Board – International Organization

Editorial Board – Global Environmental Politics

Editorial Board – Journal of Environment and Development

Grants and Fellowships

National Science Foundation, "Fostering Cross-Disciplinary Relationships and Early-Career Development to Advance Interdisciplinary Research on Climate Change and Impacts," January 2005 – December 2008, recommended for funding by the Geoscience Education program

National Science Foundation, "Analysis of the effects of environmental treaties," September 2003 - August 2006 (NSF Award No. SES-0318374)

International Studies Association Workshop Grant, 2003 (with John Duffield, Georgia State University)

American Philosophical Society, Sabbatical Fellowship, 2002-2003

Faculty Summer Research Award, Graduate School, University of Oregon, 1994, 1997, 2002

Richard A. Bray Faculty Fellow, University of Oregon, 1999

Publications

Intentional Oil Pollution at Sea: Environmental Policy and Treaty Compliance, The MIT Press, 1994, 361 pages. Winner of the Harold and Margaret Sprout Award, International Studies Association, 1995, for best book on international environmental issues.

"International Environmental Agreements: A Survey of Their Features, Formation, and Effects," *Annual Review of Environment and Resources* 28 (November 2003), 429-461.

"Knowledge Systems for Sustainable Development," David W. Cash, William C. Clark, Frank Alcock, Nancy M. Dickson, Noelle Eckley, David H. Guston, Jill Jäger, and Ronald B. Mitchell. *Proceedings of the National Academy of Sciences* 100:14 (8 July 2003), 8086-8091.

"A Quantitative Approach to Evaluating International Environmental Regimes," *Global Environmental Politics* 2:4 (November 2002), 58-83.

"Situation Structure and Institutional Design: Reciprocity, Coercion, and Exchange," Ronald B. Mitchell and Patricia M. Keilbach. *International Organization* 55:4 (Autumn 2001), 891-917.

"Implementing the Climate Change Regime's Clean Development Mechanism," Ronald B. Mitchell and Edward A. Parson. *Journal of Environment and Development* 10:2 (June 2001), 125-146.

"Discourse and Sovereignty: Interests, Science, and Morality in the Regulation of Whaling," *Global Governance* 4:3 (July-September 1998), 275-293.

"Sources of Transparency: Information Systems in International Regimes," *International Studies Quarterly* 42:1 (March 1998), 109-130.

"Empirical Research on International Environmental Policy: Designing Qualitative Case Studies" Ronald B. Mitchell and Thomas Bernauer. *Journal of Environment and Development* 7:1 (March 1998), 4-31.

"Heterogeneities at Two Levels: States, Non-State Actors, and Intentional Oil Pollution" *Journal of Theoretical Politics* 6:4 (October 1994), 625-653.

"Regime Design Matters: Intentional Oil Pollution and Treaty Compliance," *International Organization* 48:3 (Summer 1994), 425-458.

"Flexibility, Compliance and Norm Development in the Climate Regime" in <u>Implementing the Climate Regime: International Compliance</u>. Editors: Olav Schram Stokke, Jon Hovi, and Geir Ulfstein. Earthscan Press, 2005, 65-83.

"Institutions, Science, and Technology in the Transition to Sustainability," Ronald B. Mitchell and Patricia Romero Lankao, in <u>Earth System Analysis for Sustainability, Dahlem Workshop Report 91</u>. Editors: Hans Joachim Schellnhuber, Paul J. Crutzen, William C. Clark, and Martin Claussen. MIT Press, 2004, 387-407.

"A Quantitative Approach to Evaluating International Environmental Regimes" in <u>Regime Consequences: Methodological Challenges and Research Strategies</u>. Editors: Arild Underdal and Oran Young. Kluwer Academic Publishers, 2004, 121-149.

"Science, Scientists, and the Policy Process: Lessons from Global Environmental Assessments for the Northwest Forest," Ronald B. Mitchell, William C. Clark, David W. Cash, and Frank Alcock, in <u>Forest Futures: Science, Politics and Policy for the Next Century</u>. Editors: Karen Arabas and Joe Bowersox. Rowman and Littlefield, 2004, 95-111.

"Beyond Story-Telling: Designing Case Study Research in International Environmental Policy," Ronald B. Mitchell and Thomas Bernauer, in <u>Models, Numbers, and Cases: Methods for Studying International Relations</u>. Editors: Detlef Sprinz and Yael Wolinsky-Nahmias. University of Michigan Press, 2004, 81-106.

"International Environment," in <u>Handbook of International Relations</u>. Editors: Thomas Risse, Beth Simmons, and Walter Carlsnaes. Sage Publications, 2002, 500-516.

"Institutional Aspects of Implementation, Compliance, and Effectiveness" in <u>International Relations and Global Climate Change</u>. Editor: Urs Luterbacher and Detlef Sprinz. MIT Press, 2001, 221-244.

Stephen W. Pacala
Department of Ecology and Evolutionary Biology
Princeton University
Princeton, NJ 08544-1003
Tel: (609) 258-6885; Fax: (609) 258-6818

Education

B.A. 1978 Dartmouth College
Ph.D. 1982 Stanford University

Research Interests

Plant Ecology
Global Interactions of the Biosphere, Atmosphere and Hydrosphere
Mathematical Modeling
Community Ecology

Employment History

2000-present Co-Director, The Carbon Mitigation Initiative, Princeton University
1995-present Co-Director, NOAA Carbon Modeling Center, Princeton University
1994-present Associated Faculty, Princeton Environmental Institute, Princeton   University
1993-present Director of Graduate Studies, Department of Ecology and Evolutionary Biology,
              Princeton University

1992-present Professor, Department of Ecology and Evolutionary Biology, Princeton
              University
1987-1992   Associate Professor, Department of Ecology and Evolutionary Biology,
              The University Connecticut
1982-1987   Assistant Professor, Ecology Section, Biological Sciences Group, The University
              of Connecticut

1979-1981   Teaching Assistant, Stanford University
1978   Teaching Assistant, Dartmouth College
1975-1978   Research Assistant, Dartmouth College

Professional Service/Activities

Associate Editor - The American Naturalist
Associate Editor - Theoretical Population Biology
Editorial Board - Ecological Applications
Editorial Board - Global Change Biology

Publications

Kinzig, A.P. S.W. Pacala and G.D. Tilman. 2002. The Functional Consequences of Biodiversity: Experimental Progress and Theoretical Extensions. Princeton University Press, Princeton, NJ.

Hurtt, G.C., S.W. Pacala, P.R. Moorcroft, J. Caspersen, E. Shevliakova, R.A. Houghton and B. Moore III. 2002. Projecting the Future of the U.S. Carbon Sink. Proceedings of the National Academy of Sciences. 99(3):1389-1394.

Schimel, D.S., J.I. House, K.A. Hibbard, P. Bousquet, P. Ciais, P. Peylin, B.H. Braswell, M.J. Apps, D. Baker, A. Bondeau, J. Canadell, G. Churkina, W. Cramer, A.S. Denning, C.B. Field, P. Friedlingstein, C. Goodale, M. Heimann, R.A. Houghton, J.M. Melillo, B. Moore III, D. Murdiyarso, I. Noble, S.W. Pacala, I.C. Prentice, M.R. Raupach, P.J. Rayner, R.J. Scholes, W.L. Steffen and C. Wirth. 2001. Recent patterns and mechanisms of carbon exchange by terrestrial ecosystems. Nature 414:169-172.

Wilson, H.B., M.J. Keeling and S.W. Pacala. 2001. Deterministic limits to stochastic, spatial models of natural enemies. American Naturalist 159:57-80.

Moorcroft, P.R., G.C. Hurtt and S.W. Pacala. 2001. A Method for Scaling Vegetation Dynamics: the Ecosystem Demography Model (ED). Ecological Monographs 71(4):557-586.

Rees, M., R. Condit, M. Crawley, S.W. Pacala and D. Tilman. 2001. Vegetation Dynamics (9315). Science 293(5530):650-655.

Pacala S.W., Hurtt G.C., Moorcroft P.R. and Caspersen J.P. 2001. Carbon storage in the US caused by land use change. Pp. 145-172. In The Present and Future of Modeling Global Environmental Change, Terra Scientific Publishing. Toyko, Japan.

Pacala, S.W., G.C. Hurtt, R.A. Houghton, R.A. Birdsey, L. Heath, E.T. Sundquist, R.F. Stallard, D. Baker, P. Peylin, P. Moorcroft, J. Caspersen, E. Shevliakova, M.E. Harmon, S.-M. Fan, J.L. Sarmiento, C. Goodale, C.B. Field, M. Gloor and D. Schimel. 2001. Consistent Land- and Atmosphere-Based U.S. Carbon Sink Estimates. Science 292(5525):2316-2320.

Lewis, M.A. and S. Pacala. 2000. Modeling and analysis of stochastic invasion processes. Journal of Mathematical Biology 41:387-429.

Keeling, M.J., H.B. Wilson and S.W. Pacala. 2000. Re-interpreting Space, Time-lags, and Functional Responses to Ecological Models. Science 290:1758-1761.

Caspersen, J.P., S.W. Pacala, J.C. Jenkins, G.C. Hurtt, P.R. Moorcroft and R.A. Birdsey. 2000. Contributions of land-use history to carbon accumulation in US forests. Science 290:1148-1151.

Gloor, M., S.-M. Fan, S.W. Pacala and J.L. Sarmiento. 2000. Optimal sampling of the atmosphere for purpose of inverse modelling - a model study. Global Biogeochem. Cycles 14(1):407-428.

Hurtt, G.C., P.R. Moorcroft, S.W. Pacala and S. Levin. 1998. Terrestrial Models and Global Change: Challenges for the Future. Global Change Biology 4(5):581-590.

Diane E. Pataki
Dept. of Earth System Science
University of California, Irvine
Irvine, CA 92697-3100
Tel. (949) 824-9411

Education

1998   Ph.D.   Nicholas School of the Environment, Duke University, Durham, NC
1995   M.S.   Nicholas School of the Environment, Duke University, Durham, NC
1993   B.A.   Barnard College, Columbia University, New York, NY

Employment History

2004-   Assistant Professor. Dept. of Earth System Science and Dept. of Ecology & Evolutionary Biology, University of California, Irvine
2004-   Adjunct Assistant Professor. Department of Biology, University of Utah
2000-2004   Research Assistant Professor. Department of Biology, University of Utah
1999-2003   IGBP-GCTE Focus 1 Scientific Officer. University of Utah
1998-1999   Post-doctoral Research Associate. Desert Research Institute

Professional Service/Activities

Board of Advisors to the Editor, *New Phytologist* journal
Steering Committee, Biosphere-Atmosphere Stable Isotope Network (BASIN)
Steering Committee, Terrestrial Ecosystem Responses to Atmospheric and Climatic Change (TERACC) Network
NSF Ecosystems panelist, spring 2005
Ecological Society of America
American Geophysical Union
International Association for Urban Climate

Publications

Pataki DE, Bush SE, Ehleringer JR. 2005. Stable isotopes as a tool in urban ecology. *In* Stable isotopes and biosphere-atmosphere interactions: Processes and biological controls. Flanagan LB, Ehleringer JR, Pataki DE, Eds. Elsevier Press, San Diego, pp 199-214.

Flanagan LB, Ehleringer JR, Pataki DE, Eds. 2005. Stable isotopes and biosphere-atmosphere interactions: Processes and biological controls. Elsevier Press, San Diego.

Luo Y, Su B, Currie WS, Dukes JS, Finzi A, Hartwig U, Hungate B, McMurtrie R, Oren R, Parton WJ, Pataki DE, Shaw R, Zak DR, Field C. 2004. Progressive nitrogen limitation of ecosystem responses to rising atmospheric CO_2. Bioscience 54(8): 731-739.

Morgan JA, Pataki DE, Gruenzweig J, Körner C, Newton P, Niklaus PA, Nippert J, Nowak RS, Parton W, Clark H, Del Grosso SJ, Knapp AK, Mosier AR, Polley W, Shaw R. 2004. Grassland responses to rising atmospheric CO_2 are driven primarily by water relations. Oecologia 140: 11-25.

Pataki DE, Bowling DR, Ehleringer JR. 2003. Seasonal cycle of carbon dioxide and its isotopic composition in an urban atmosphere: anthropogenic and biogenic effects. Journal of Geophysical Research – Atmospheres 108(D23), 4735

Pataki DE, Ellsworth DW, King JS, Leavitt SW, Lin G, Pendall E, Siegwolf R, van Kessel C, Ehleringer JR. 2003. Tracing changes in ecosystem function under elevated CO_2. Bioscience 53(9): 805-818.

Bowling DR, Pataki DE, Ehleringer JR. 2003. Critical evaluation of micrometeorological methods for measuring ecosystem-atmosphere isotopic exchange of CO_2. Agricultural and Forest Meteorology 116: 159-179.

Pataki DE, Ehleringer JR, Flanagan LB, Yakir D, Bowling DR, Still C, Buchmann N, Kaplan JO, Berry JA. 2003. The application and interpretation of Keeling plots in terrestrial carbon cycle research. Global Biogeochemical Cycles 17(1), 1022

Canadell J, Pataki DE. 2002. New advances in carbon cycle research. Trends in Ecology and Evolution 17(4): 156-158.

Pataki DE. 2002. Atmospheric CO_2, climate and evolution – lessons from the past. New Phytologist 154:10-14.

Pataki DE, Huxman TE, Jordan DN, Zitzer SF, Coleman JS, Smith SD, Nowak RS, Seemann JR. 2000. Water use of two Mojave Desert shrubs under elevated CO_2. Global Change Biology 6(8): 889-898.

Pataki, D.E., R. Oren, and D.T. Tissue. 1998. Elevated carbon dioxide does not affect stomatal conductance of *Pinus taeda* L. *Oecologia* 117: 47-52.

Keith H. Paustian
Natural Resource Ecology Lab
Colorado State University
Ft. Collins, CO 80523
Tel: (970) 491-1547; Fax: (970) 491-1965

Education

1977 B.Sc. Forest Biology, Colorado State University, Fort Collins
1980 M. Sc. Forest Ecology, Colorado State University, Fort Collins
1987 Ph.D. Systems Ecology/Agroecology, Swedish University of Agricultural
              Sciences, Uppsala

Research Interests

Soil carbon sequestration in grasslands; mechanisms of soil carbon storage; modeling the carbon cycle in managed ecosystems.

Employment History

2001-present Professor, Department of Soil and Crop Sciences, Colorado State University
1996-present Senior Research Scientist, Natural Resource Ecology Lab, Colorado State University
1993-1995 Research Scientist, Natural Resource Ecology Lab, Colorado State University
1991-1993 Research Assistant Professor, W.K. Kellogg Biological Station, Michigan State University
1989-1990 Research Associate, W.K. Kellogg Biological Station, Michigan State University
1987-1989 Research Scientist, Dept. of Ecology and Environmental Research, Swedish University of Agricultural Sciences

Professional Service/Activities

Executive Committee – Consortium for Agricultural Mitigation of Greenhouse Gases (CASMGS)
Coordinating Lead Author – IPCC Good Practice Guidelines for Land Use, Land Use Change and Forestry, National Inventory Guidelines

Lead Author - IPCC Special Report on a "Land use, Land use Change and Forestry" Review Team - New Zealand National Carbon Inventory System

Co-chair for CAST Taskforce on climate change impacts and greenhouse gas mitigation in U.S. agriculture

Planning Committee member - Terrestrial Ecosystems Research Facilities, Department of Energy

Steering Committee member - International Geosphere Biosphere Program/Global Change in
Terrestrial Ecosystems, Focus 3, Soil Organic Matter
Steering Committee Member – U.S. Climate Change National Assessment, Agricultural
Sector Team
Task Force member - DOE National Taskforce to develop a Carbon Sequestration Roadmap

Co-chair - IPCC Working Group on Methodologies for Country Inventories of
Greenhouse Gases: CO_2 Emissions from soils

Publications

Paustian, K., E.T. Elliott, G.A. Peterson and K. Killian. 1996. Modelling climate, CO_2 and
management impacts on soil carbon in semi-arid agroecosystems. Plant and Soil 187:351-
365.

Paustian, K., O. Andren, H. Janzen, R. Lal, P. Smith, G. Tian, H. Tiessen, M. van Noordwijk and
P. Woomer. 1997. Agricultural soil as a C sink to offset CO_2 emissions. Soil Use and
Management 13:230-244.

Paustian, K., C.V. Cole, D. Sauerbeck and N. Sampson. 1998. CO_2 mitigation by agriculture: An
overview. Climatic Change 40:135-162.

Paustian, K., E.T. Elliott, J. Six and H.W. Hunt. 2000. Management options for reducing CO_2
emissions from agricultural soils. Biogeochemistry 48:147-163.

Collins, H.P., E.T. Elliott, K. Paustian, L.G. Bundy, W.A. Dick, D.R. Huggins, A.J.M. Smucker
and E.A. Paul. 2000. Soil carbon pools and fluxes in long-term Corn Belt
agroecosystems. Soil Biol. Biochem. 32:157-168.

Paustian, K., E.T. Elliott, K. Killian, J. Cipra, G. Bluhm and J.L. Smith. 2001. Modeling and
regional assessment of soil carbon: A case study of the Conservation Reserve Program.
In: R. Lal and K. McSweeney (eds) Soil Management for Enhancing Carbon
Sequestration. Pp. 207-225. SSSA Special Publ., Madison, WI.

Conant, R.T., K. Paustian and E.T. Elliott. 2001. Grassland management and conversion into
grassland: Effects on soil carbon. Ecological Application 11:343-355.

Eve, M.D., M. Sperow, K. Paustian and R.F. Follett. 2002. National-scale estimation of changes
in soil carbon stocks on agricultural lands. Environmental Pollution 116: 431-438.

Conant, R.T. and K. Paustian 2002. Potential soil carbon sequestration in overgrazed grassland
ecosystems. Global Biogeochemical Cycles 16:90_1-90_9.

Eve, M.D., M. Sperow, K. Howerton, K. Paustian and R.F. Follett. 2002. Predicted impact of
management changes on soil carbon stocks for each cropland region of the conterminous
U.S. Journal of Soil and Water Conservation 57:196-204.

Antle, J.M., S.M. Capalbo, S. Mooney, E. Elliott and K. Paustian. 2002. Economic Analysis of
Agricultural Soil Carbon Sequestration: An Integrated Assessment Approach. Journal of
Agricultural and Resource Economics 26:344-367.

Antle, J.M., S.M. Capalbo, S. Mooney, E.T. Elliott and K. H. Paustian. 2002. A comparative
examination of the efficiency of sequestering carbon in U.S. agricultural soils. American
Journal of Alternative Agriculture 17:109-115.

Reilly, J., F. Tubiello, B. McCarl, D. Abler, R. Darwin, K. Fuglie, S. Hollinger C. Izaurralde, S.
Jagtap, J. Jones, L. Mearns, D. Ojima, E. Paul, K. Paustian, S. Riha, N. Rosenberg, C.
Rosenzweig. 2003. U.S. Agriculture and Climate Change: New Results. Climatic
Change 57:43-69.

Sperow, M., M.D. Eve and K. Paustian. 2003. Potential soil C sequestration on U.S. agricultural soils. Climatic Change 57:319-339.

DeGryze, S., J. Six, K. Paustian, S.J. Morris, E.A. Paul and R. Merckx. 2003. Soil organic carbon pool changes following land use conversions. Global Change Biology (in press).

Ogle, S.M., R.T. Conant and K. Paustian. 2003. Deriving grassland management factors for a carbon accounting method developed by the Intergovernmental Panel on Climate Change. Environ. Management (in press).

Conant, R.T. and K. Paustian. 2003. Grassland management activity data: current sources and future needs. Environ. Management (in press).

Paustian, K. and B. Babcock (eds). 2003. Climate Change and Greenhouse Gas Mitigation: Challenges and Opportunities for Agriculture. Council on Agricultural Sciences and Technology (CAST). (In press).

Jorge L. Sarmiento
AOS Program, Princeton University
Sayre Hall, Forrestal Campus
P.O. Box CN710
Princeton, NJ 08544-0710
Tel. (609): 258-6585; Fax: (609) 258-2850

Education

1968 B.A. in Chemistry	Swarthmore College
1974 M.A. in Geology	Columbia University
1976 M.Ph. in Geology	Columbia University
1978 Ph.D. in Geology	Columbia University

Employment History

1973-1978 Graduate Research Assistant, Columbia University
1978-1980 Research Associate in Atmospheric and Oceanic Sciences Program, Princeton
           University
1980-1986 Assistant Professor in Geological and Geophysical Sciences Department,
           Atmospheric and Oceanic Sciences Program, Princeton University
1986-1991 Associate Professor in Geological and Geophysical Sciences Department,
           Atmospheric and Oceanic Sciences Program, Princeton University
1991-present   Professor in Geological and Geophysical Sciences Department, Atmospheric
           and Oceanic Sciences Program, Princeton University
1995-present   Associated Faculty in Department of Civil Engineering and Operations
           Research, Princeton University
1996-present   Associated Faculty in Princeton Environmental Institute, Princeton
           University

Professional Service/Activities

1995-2003	Member U.S. JGOFS ExecPlus Committee
1995-2003	Co-Chairman U.S. JGOFS Synthesis and Modeling Project
1998-1999	Co-Chairman Carbon and Climate Planning Group, USGCRP
2000-	Member-U.S. Carbon Scientific Steering Group, USGCRP
2003-	Director, NOAA/Princeton Cooperative Institute on Climate Science
Summer 1993	H. Burr Steinbach Visiting Scholar, Woods Hole Oceanographic Institution
1994-1995	Visiting Professor, Physikalisches Institut, Universität Bern, Bern, Switzerland
1998-1999	Bourse a haut-Niveau from the French Minister of Science
2003	Fellow of the American Geophysical Union Fellow
2004	Fellow of the American Association for the Advancement of Science

American Association for the Advancement of Science
American Geophysical Union
American Meteorological Society
American Society of Limnology and Oceanography
Oceanography Society
Sigma Xi

Publications

Gloor, M., N. Gruber, J. L. Sarmiento, C. S. Sabine, R. Feely, and C. Rödenbeck, 2003. A first estimate of present and pre-industrial air-sea CO_2 flux patterns based on ocean carbon measurements. Geophys. Res. Lett., 30(1): 10.1029/2002GL015594.

McNeil, B. I., R. J. Matear, R. M. Key, J. L. Bullister, and J. L. Sarmiento, 2003. Anthropogenic CO_2 uptake by the ocean based on the global chlorofluorocarbon dataset Science, 299: 235-239.

Toggweiler, J. R., A. Gnanadesikan, S. Carson, R. Murnane, and J. L. Sarmiento, 2003. Representation of the carbon cycle in box models and GCMs, Part 1, the solubility pump, Global Biogeochem. Cycles, 17(1): 1026, doi:10.1029/2001GB001401.

Toggweiler, J. R., R. Murnane, S. Carson, A. Gnanadesikan, and J. L. Sarmiento, 2003. Representation of the carbon cycle in box models and GCMs, Part 2, the organic carbon pump, Global Biogeochem. Cycles, 17(1): 1027, doi:10.1029/2001GB001841.

Gnanadesikan, A., J. L. Sarmiento, and R. D. Slater, 2003. Effects of patchy ocean fertilization on atmospheric carbon dioxide and biological production. Global Biogeochem. Cycles, 17 (2), doi: 10.1029/2002GB001940.

Law, R. M., Y.-H. Chen, K. R. Gurney, and TransCom 3 modellers, 2003. TransCom 3 CO_2 inversion intercomparison: 2. Sensitivity of annual mean results to data choices. Tellus, 55B (2): 580-595.

Gurney, K. R., R. M. Law, et al., 2003. TransCom 3 CO_2 inversion intercomparison: 1. Annual mean control results and sensitivity to transport and prior flux information. Tellus, 55B (2): 555-579.

Patra, P. K., S. Maksyutov, and TransCom-3 Modelers, 2003. Sensitivity of optimal extension of observation networks to the model transport. Tellus, 55B (2): 498-511.

Maksyutov, S., T. Machida, H. Mukai, P. Patra, T. Nakazawa, G. Inoue, and TransCom-3 Modelers, 2003. Effect of recent observations on Asian CO2 flux estimates with transport model inversions. Tellus 55B (2): 522-529.

Gao, Y., S.-M. Fan, and J. L. Sarmiento, 2003. Aeolian iron input to the ocean through precipitation scavenging: a modeling perspective and its implication for natural iron fertilization in the ocean. (J. Geophys. Res., 108(D7), 4221, doi:10.1029/2002JD002420.

Watson, A. J., J. C. Orr, et al. 2003. Carbon dioxide fluxes in the global ocean. In: Ocean Biogeochemistry, ed. M. J. R. Fasham, Springer-Verlag, Publishers, New York, pp. 123-143.

Sarmiento, J. L., N. Gruber, M. A. Brzezinski, and J. P. Dunne, 2004. High latitude controls of the global nutricline and low latitude biological productivity. Nature, 427: 56-60.

Matsumoto, K., J.L. Sarmiento, et al., 2004. Evaluation of ocean carbon cycle models with data-based metrics. Geophys. Res. Lett., 31, L07303, doi:10.1029/2003GL018970.

Mignone, B. K., J. L. Sarmiento, R. D. Slater, and A. Gnanadesikan, 2004. Sensitivity of sequestration efficiency to mixing processes in the global ocean. Energy, 29: 1467-1478

Greenblatt, J. B., and J. L. Sarmiento, 2004. Variability and climate feedback mechanisms in ocean uptake of CO_2. In: The Global Carbon Cycle, ed. C. B. Field and M. R. Raupach, Island Press, Washington, D.C., pp. 257-275.

Edmonds, J., F. Joos, N. Nakicenovic, R. G. Richels, and J. L. Sarmiento, 2004. Scenarios, targets, gaps, and costs. In: The Global Carbon Cycle, ed. C. B. Field and M. R. Raupach, Island Press, Washington, D.C., pp. 77-102.

Sarmiento, J. L., R. Slater, R. Barber, L. Bopp, S. C. Doney, A. C. Hirst, J. Kleypas, R. Matear, U. Mikolajewicz, P. Monfray, V. Soldatov, S. A. Spall, and R. Stouffer, 2004. Response of ocean ecosystems to climate warming. Global Biogeochem . Cycles, 18, GB3003, doi:1029/2003GB002134.

Marinov, I., and J. L. Sarmiento, 2004. The role of the oceans in the global carbon cycle: An overview. In: The Ocean Carbon Cycle and Climate, ed. M. Follows and T. Oguz, NATO ASI, Ankara, Turkey, Kluwer Academic Publishers, pp. 251-295.

Doney, S. C., K. Lindsay, et al., 2004. Evaluating global ocean carbon models: The importance of realistic physics, Global Biogeochem. Cycles, 18, GB3017, doi:10.1029/2003GB002150.

Gnanadesikan, A., J. P. Dunne, R. M. Key, K. Matsumoto, J. L. Sarmiento, R. D. Slater, and P, S. Swathi, 2004. Oceanic ventilation and biogeochemical cycling: Understanding the physical mechanisms that produce realistic distributions of tracers and productivity. Global Biogeochem. Cycles, 18, GB4010, doi:10.1029/2003GB002097.

Taro Takahashi
Lamont-Doherty Earth Observatory
Columbia University
61 Route 9W
Palisades, NY 10964-8000
Tel. (845) 365-8537

Education:
 Bachelor of Engineering, University of Tokyo, JAPAN, 1953
 Ph. D. (Earth Science), Columbia University, New York, NY, 1957

Positions Held:
 Doherty Senior Scholar, Lamont-Doherty Earth Observatory, Columbia
  University, Palisades, NY, 1998-present.
 Adjunct Professor, Department of Earth and Environmental Sciences, Columbia
  University, New York, NY, 1978-present.
 Associate Director, Lamont-Doherty Earth Observatory, Columbia University, Palisades,
  NY, 1981-2003.
 Acting Chief Executive Officer, Biosphere-2 Center Inc., An affiliate of Columbia
  University, Oracle, AZ, Jan.,-June, 1996.
 Doherty Senior Scientist, Lamont-Doherty Earth Observatory, Columbia University,
  Palisades, NY, 1984-1998.
 Senior Research Scientist, Lamont-Doherty Geological Observatory, Columbia
  University, Palisades, NY, 1977-1984
 Distinguished Professor of Physical Sciences, Queens College, City University of New
  York, Queens, NY, 1971-1977
 Visiting Associate, California Institute of Technology, Pasadena, CA, 1970-1971
 Professor, University of Rochester, Rochester, NY, 1969-1970

Recent Professional Services:
 Chairman, International Steering Committee for Ocean Nourishment in Asia, 1998-2003.
 Member, Ocean CO_2 Panel, CCCO/UNESCO, Paris, 1988-2000.

 Member, Advisory panel on the Roadmap for the Science and Technology for Carbon
  Management, Department of Energy, Germantown, VA, October, 1998.
 Member, Scientific Steering Committee, International JGOFS Program for the Joint Global
  Ocean Flux Study (JGOFS) Core Project, International Geosphere-Biosphere Program
  (IGBP) and the Scientific Committee on Oceanic Research (SCOR), 1995-1997.
 Member, Science Steering Committee, US JGOFS Program, June, 1994-1997.
 Member and co-chairman, Advisory Panel for the Oceanic CO_2 Flux Program, National
  Oceanographic and Atmospheric Administration, Washington, D. C., 1988-1997.
 Member, Scientific Steering Committee, U. S. JGOFS, 1994-1997.
 Chairman, Committee on Ocean CO_2, Ocean Study Board, National Academy of
  Sciences/National Research Council, Washington, D. C., 1992-1995.
 Secretary, The Geochemical Society, 1980-1983

Co-chairman, 4th Ewing Symposium on the Climate Processes and Climate Sensitivity, 1982.

Honors Received:
Ford Prize, Ford Motor Company, Dearborn, MI, 2004.
Fellow, American Geophysical Union, 2003
Outstanding Paper Award, Office of Oceanic and Atmospheric Research, NOAA, 2000.
Distinguished Authorship Award (shared with P. P. Tans and I. Fung), NOAA, Washington, D. C. 1991

Selected Publications:
Books:
"Climate Processes and Climate Sensitivity" (1984) edited by J. E. Hansen and T. Takahashi, Geophysical Monograph #29, American Geophysical Union, Washington, D.C., 368 pp.

Professional Journal Articles:

Hales, B., Takahashi, T. and Bandstra, L. (2005). Atmospheric CO_2 uptake by a coastal upwelling system. Global Biogeochem. Cycles, 19. doi.10.1029/2004GB002295.

Hales, B. and Takahashi, T. (2004). High-resolution biogeochemical investigation of the Ross Sea Antarctica, during the AESOPS (U. S.JGOFS) Program . Global Biogeochem. Cycles, Vol. 18, No. 3, GB3006, doi. 10.1029/2003GB002165.

K. H. Coale, K. S. Johnson and the members of the SOFeX Project. (2004). Southern Ocean iron enrichment experiment: Carbon cycling in high- and low-Si waters. Science, 304, 408-414.

Takahashi, T. (2004) Fate of industrial carbon dioxide, Science, 305, 352-353.

Hales, B., Chipman, D. W. and Takahashi, T. (2004). High-frequency measurement of partial pressure and total concentration of carbon dioxide in seawater using microporous hydrophobic membrane contactors. Limnol. & Oceanogr. Methods, 2, 356-364.
Millero, F. J., Pierrot, D. Lee, K., Wanninkhof, R., Feely, R., Sabine, C. L., Key, R. M and Takahashi, T. (2002). Dissociation constants for carbonic acid determined from field measurements. Deep-Sea Res., 49, 1705-1724.

Hales, B. and Takahashi, T. (2002). The pumping SeaSoar: A high-resolution seawater sampling platform. Jour. Oceanic and Atmospheric Technology 19, 1096-1104.

Takahashi, T., Sutherland, S. C., Feely, R. A. and Cosca, C. (2003). Decadal variation of the surface water pCO_2 in the western and central Equatorial Pacific. Science, 302, 852-856.

Hales, B., van Geen, A. and Takahashi, T. (2004). High-frequency measurement of seawater chemistry: Flow-injection analysis of macronutrients. Limnol. Oceanogr. Methods (2004) 2:91-101.

Takahashi, T., Sutherland, S. C., Sweeney, C., Poisson, A., Metzl, N., Tillbrook, B., Bates, N., Wanninkhof, R., Feely, R. A., Sabine, C., Olafsson, J. and Nojiri, Y. (2002). Global sea-air CO_2 flux based on climatological surface ocean pCO_2, and seasonal biological and temperature effects, Deep-Sea Res. II, 49, 1601-1622.

Rubin, S. I., Takahashi, T., Chipman, D. W. and Goddard, J. G. (1998). Primary production and nutrient utilization ratios in the Pacific sector of the Southern Ocean based on seasonal changes in seawater chemistry. Deep-Sea Res., 45, Part I, 1211-1234.

Takahashi, T., Feely, R. A., Weiss, R., Wanninkhof, R. H., Chipman, D. W., Sutherland, S. C. and Takahashi, T. T. (1997). Global air-sea flux of CO_2: an estimate based on measurements of sea-air pCO_2 difference. Proc. National Acad. Sci., 94, 8292-8299.

Takahashi, T., Takahashi, T. T. and Sutherland, S. C. (1995). An assessment of the role of the North Atlantic as a CO_2 sink. Phil. Trans. Roy. Soc. London, Series B, 348, 143-152.

Chipman, D. W., Marra, J. and Takahashi, T. (1993). Primary production at 47°N and 20°W in the North Atlantic Ocean: A comparison between the ^{14}C incubation method and the mixed layer carbon budget. Deep-Sea. Res., 40, 151-170.

Takahashi, T., Olafsson, J., Goddard, J., Chipman, D. W. and Sutherland, S. C., (1993). Seasonal variation of CO_2 and nutrients in the high-latitude surface oceans: A comparative study. Global Biogeochemical Cycles, 7, 843-878.

Tans, P. P., Fung, I. Y. and Takahashi, T. (1990). Observational constraints on the global atmospheric CO_2 budget. Science, 247, 1431-1438.

Pieter P. Tans
Environmental Research Laboratories
Climate Monitoring and Diagnostics Laboratory (CMDL)
National Oceanic and Atmospheric Administration (NOAA)
325 Broadway, Boulder, CO 80303-3328
Tel: (303) 497-6811

Education

1973   Doctorandus, Theoretical Physics (cum laude)
1978   Ph.D. , Experimental Physics, Rijkusuniversiteit Groningen, The Netherlands

Research Interests

Past research: Magnetic impurities in an electron lattice gas; One-dimensional radiative climate model; High precision ^{14}C counting; Stable isotopes in tree rings; Radioisotope detection with a cyclotron; Development of Raman scattering method to detect minute changes in the ratio of atmospheric O_2 to N_2.

Present: Biogeochemical cycles; Global climate change; Stable isotope applications; Atmospheric chemistry and transport; Inverse models; Air-sea exchange of gases; Development of new generation of accurate and robust gas analyzers

Employment History

1978-1979   Postdoc, Scripps Inst. Oceanography, La Jolla, CA, with C.D. Keeling.
1979-1985   Staff scientist, Astrophysics Group, Lawrence Berkeley Laboratory, Berkeley.
1985-1990   Research Associate, CIRES, University of Colorado, Boulder.
1990-1996   Supervisory Physicist, Climate Monitoring and Diagnostics Laboratory, National Oceanic and Atmospheric Administration, Boulder.
1996-present   Chief Scientist, Climate Monitoring and Diagnostics Laboratory.

Professional Service/Activities

1992-2000   Department of Chemistry & Biochemistry, University of Colorado at Boulder.
1992-1993   Committee on Oceanic Carbon, Ocean Studies Board, NRC
1995-1997   Dec-Cen Panel, Board on Atmospheric Sciences and Climate, NRC
1997-Present   CIRES fellow
1998-1998   Working Group drafting a multi-agency U.S. Carbon Cycle Science Plan
1996-Present   Associate Editor, Journal of Climate
1997-Present   Editorial Advisory Board, Tellus B
1995-Present   Corresponding member, Royal Dutch Academy of Sciences
2000-Present   Gold Medal, Department of Commerce
2002-Present   ISI Highly Cited (248 most cited authors in the geosciences 1981-1999)

2004-Present Fellow, American Geophysical Union

Publications

Francey, R. J. and P. P. Tans, Latitudinal variation in oxygen-18 of atmospheric CO_2, **Nature 327**, 495-497, 1987.

Tans, P. P., T. J. Conway, and T. Nakazawa, Latitudinal distribution of the sources and sinks of atmospheric carbon dioxide derived from surface observations and atmospheric transport model, **J. Geophys. Res. 94**, 5151-5172, 1989.

Tans, P. P., I. Y. Fung, and T. Takahashi, Observational constraints on the global atmospheric carbon dioxide budget, **Science 247**, 1431-1438, 1990.

Steele, L.P., E.J. Dlugokencky, P.M. Lang, P.P. Tans, R.C. Martin, and K.A. Masarie, Slowing down of the global accumulation of atmospheric methane during the 1980's, **Nature, 358**, 313-316, 1992.

Tans, P.P., J.A. Berry, and R.F. Keeling, Oceanic $^{13}C/^{12}C$ observations, a new window on CO_2 uptake by the oceans, **Glob. Biogeochem. Cycles, 7,** 353-368, 1993.

Novelli, P.C., K.A. Masarie, P.P. Tans, and P.M. Lang, Recent changes in atmospheric carbon monoxide, **Science, 263,** 1587-1590, 1994.

Bender, M.L., P.P. Tans, J.T. Ellis, J. Orchardo, and K. Habfast, A high precision isotope ratio mass spectrometry method for measuring the O_2/N_2 ratio of air, **Geochim. Cosmochim. Acta, 58,** 4751-4758, 1994.

Ciais, P., P.P. Tans, M. Trolier, J.W.C. White, and R.J. Francey, A large northern hemisphere terrestrial CO_2 sink indicated by the $^{13}C/^{12}C$ ratio of atmospheric CO_2, **Science, 269,** 1098-1102, 1995.

Battle, M., M. Bender, T. Sowers, P. Tans, J. Butler, J. Elkins, J. Ellis, T. Conway, N. Zhang, P. Lang, and A. Clarke, Atmospheric gas concentrations over the past century measured in air from firn at the South Pole, **Nature 383,** 231-235, 1996.

Tans, P.P., Why carbon dioxide from fossil fuel burning won't go away, in: Perspectives in Environmental Chemistry, edited by D. Macalady, Ch. 12, pp. 271-291, Oxford University Press, New York, 1998.

Tans, P.P., The CO_2 lifetime concept should be banished, **Climatic Change,** 37, 487-490, 1997.

Fan, S., M. Gloor, J. Mahlman, S. Pacala, J. Sarmiento, T. Takahashi, and P. Tans, A large terrestrial sink in North America implied by atmospheric and oceanic carbon dioxide data and models, **Science 282**, 442-446, 1998.

Bousquet, Philippe, Philippe Peylin, Philippe Ciais, Corinne le Quere, Pierre Friedlingstein, and Pieter Tans, Regional changes in carbon dioxide fluxes of land and oceans since 1980, **Science 290**, 1342-1346, 2000.

Charles Tarnocai, M.Sc.
Research Scientist
Agriculture and Agri-Food Canada
Research Branch
OTTAWA, Canada, K1A 0C6
Tel.: (613) 759-1857; Fax: (613) 759-1937

Areas of Responsibility:
- Soil genesis and soil taxonomy
- Soil resource-mapping
- Wetlands and peatlands
- Soil carbon and climate change
- Cryogenic or permafrost-affected soils
- Organic soils (Histosols)
- Paleosols and paleoclimate
- Soil database systems-soil and peatland carbon, circumpolar, multilayer
- Environmental monitoring

Publications since 2000 (of 260 total)

Tarnocai, C., I.M. Kettles and B. Lacelle. 2000. Peatlands of Canada. Geological Survey of Canada, Ottawa, Open File 3834. (map)

Tarnocai, C., I.M. Kettles and B. Lacelle. 2000. Peatlands of Canada digital database. Geological Survey of Canada, Ottawa, Open File 3834. (database)

Tarnocai, C. 2000. Carbon pools in soils of the Arctic, Subarctic and Boreal regions of Canada. In: R. Lal, J.M. Kimble and B.A. Stewart (eds.), *Global Climate Change and Cold Regions Ecosystems.* Advances in Soil Science, Lewis Publishers, Boca Raton, Fla., pp. 91–103.

Tarnocai, C. 2000. Analysis of the Pony Creek Core (DDH-115). In: L.E. Jackson. Quaternary Geology of the Carmacks Map Area, Yukon Territory. Geological Survey of Canada Bulletin 539. Natural Resources of Canada, Ottawa. p.52–56.

Swanson, D.K, B. Lacelle and C. Tarnocai. 2000. Temperature and the Boreal-Subarctic maximum in soil organic carbon. *Géographie physique et Quaternaire*, 54(2):153–163.

Desjardins, R.L., W. Smith, B. Grant, C. Tarnocai and J. Dumanski. 2001. Soil Carbon Sequestration and Greenhouse Effect, Soil Science Society of America Special Publication No. 57. L. Lal, editor. Madison, WI, USA. p. 115–123.

Tarnocai, C. 2001. Cryosols. p. 277–283 in: P. Driessen, J. Deckers, O. Spaargaren and F. Nachtergaele (eds.), *Major Soils of the World*, World Soil Resources No. 94, Food and Agriculture Organization of the United Nations, Rome, Italy.

Tarnocai, C. 2001. Wetlands of Canada. Prepared for Climate Change Action Fund, Agriculture and Agri-Food Canada, Research Branch, ECORC. 14 p

Tarnocai, C. and B. Lacelle. 2001. Wetlands of Canada Database. Agriculture and Agri-Food Canada, Research Branch, ECORC, NSDB. (digital database)

Tarnocai, C., I. Kettles and B. Lacelle. 2001. Wetlands of Canada. Agriculture and Agri-Food Canada, Research Branch. (map)

Tarnocai, C., J. Kimble, D. Swanson, S. Goryachkin, Ye. Naumov, V. Stolbovoi, B. Jakobsen, G. Broll, L. Montanarella, O. Arnalds, A. Arnoldussen, F. Orozco-Chavez and M. Yli-Halla. 2001. Soils of Northern and Mid Latitudes. Research Branch, Agriculture and Agri-Food Canada, Ottawa, Canada. (map)

Tarnocai, C., J. Kimble, D. Swanson, S. Goryachkin, Ye. Naumov, V. Stolbovoi, B. Jakobsen, G. Broll, L. Montanarella, A. Arnoldussen, O. Arnalds and M. Yli-Halla. 2001. Northern Circumpolar Soils. Research Branch, Agriculture and Agri-Food Canada, Ottawa, Canada. (map)

Tarnocai, C., J. Gould, G. Broll and P. Achuff. 2001. Ecosystem and Trafficability Monitoring for Quttinirpaaq National Park. Prepared for Parks Canada. Agriculture and Agri-Food Canada, Research Branch, ECORC. 169 p.

Bhatti, J.S., M.J. Apps and C. Tarnocai. 2002. Estimates of soil organic carbon stocks in central Canada using three different approaches. *Canadian Journal of Forest Research*, 32:805–812.

Tarnocai, C. and I. Campbell. 2002. Soils of the Polar Regions. In: *Encyclopedia of Soil Science*, Marcel Dekker, Inc., New York, U.S.A. pp.1018–1021.

Tarnocai, C. 2002. Cryosols in a changing environment: their role and research needs. Transactions of the 17th World Congress of Soil Science, August 14–21, 2002, Bangkok, Thailand, Paper 86, CD-ROM.

Broll, G. and C. Tarnocai. 2002. Turf hummocks on Ellesmere Island, Canada. Transactions of the 17th World Congress of Soil Science, August 14–21, 2002, Bangkok, Thailand, Paper 1049, CD-ROM.

Goryachkin, S., V. Stolbovoi, C. Tarnocai, J. Kimble, G. Broll, B. Jakobsen, L. Montanarella, E. Naumov, A. Arnoldussen, B. Lacelle and S. Waltman. 2002. Northern circumpolar soil database and derived soil maps in different classification systems. Transactions of the 17th World Congress of Soil Science, August 14–21, 2002, Bangkok, Thailand, Paper 838, CD-ROM.

Tarnocai, C., J. Kimble, D. Swanson, S. Goryachkin, Ye. Naumov, V. Stolbovoi, B. Jakobsen, G. Broll, L. Montanarella, O. Arnalds, A. Arnoldussen, F. Orozco-Chavez and M. Yli-Halla. 2002. Soils of Northern and Mid Latitudes. Version 1. Research Branch, Agriculture and Agri-Food Canada, Ottawa, Canada. (map)

Tarnocai, C., J. Kimble, D. Swanson, S. Goryachkin, Ye. Naumov, V. Stolbovoi, B. Jakobsen, G. Broll, L. Montanarella, A. Arnoldussen, O. Arnalds and M. Yli-Halla. 2002. Northern Circumpolar Soils. Version 1. Research Branch, Agriculture and Agri-Food Canada, Ottawa, Canada. (map)

Nixon, M., C. Tarnocai and L. Kutny. 2003. Long-term active layer monitoring: Mackenzie Valley, northwest Canada. In M. Philips, S. Springman and L.U. Arenson (eds.), *Permafrost*, Vol. 2, A.A. Balkema Publishers, Swets & Zeitlinger, Lisse, The Netherlands, pp. 821–826.

Tarnocai, C., J. Kimble and G. Broll. 2003. Determining carbon stocks in Cryosols using the Northern and Mid Latitudes Soil Database. In M. Philips, S. Springman and L.U. Arenson (eds.), *Permafrost*, Vol. 2, A.A. Balkema Publishers, Swets & Zeitlinger, Lisse, The Netherlands, ISBN 90 5809 582 7, Volume 2, pp. 1129–1134.

Broll, G., C. Tarnocai and J. Gould. 2003. Long-term High Arctic ecosystem monitoring in Quttinirpaaq National Park, Ellesmere Island, Canada. In M. Philips, S. Springman and L.U. Arenson (eds.), *Permafrost*, Vol. 2, A.A. Balkema Publishers, Swets & Zeitlinger, Lisse, The Netherlands, ISBN 90 5809 582 7, Volume 1, pp. 89–94.

Tarnocai, C. 2003. Sensitivity of Canadian peatlands to climate change. Mitteilungen Bodenkundlichen Gesellschaft, Vol. 101, p. 135–136.

Tarnocai, C. 2003. The effect of climate change on northern agriculture in Canada. Mitteilungen Bodenkundlichen Gesellschaft, Vol. 101, p. 139–140.

Tarnocai, C., J. Kimble, S. Waltman, D. Swanson, S. Goryachkin, Ye. Naumov, V. Stolbovoi, B. Jakobsen, G. Broll, L. Montanarella, O. Arnalds, A. Arnoldussen, F. Orozco-Chavez and M. Yli-Halla. 2003. Soils of Northern and Mid Latitudes. Research Branch, Agriculture and Agri-Food Canada, Ottawa, Canada. (1:15 000 000 scale map)

Tarnocai, C., J. Kimble, S. Waltman, D. Swanson, S. Goryachkin, Ye. Naumov, V. Stolbovoi, B. Jakobsen, G. Broll, L. Montanarella, A. Arnoldussen, O. Arnalds and M. Yli-Halla. 2003. Northern Circumpolar Soils. Research Branch, Agriculture and Agri-Food Canada, Ottawa, Canada. (1:10 000 000 scale map)

Tarnocai, C., Kettles, I. M. and Lacelle, B. 2003. Peatlands of the Mackenzie River Valley. (Database and map). Geological Survey of Canada, Ottawa, Open File 4413. Ottawa, Canada.

Tarnocai, C. 2004. Organic Carbon in Cryosolic Soils in the Northern Circumpolar Area. CliC Newsletter "Ice and Climate", p. 10-11.

Tarnocai, C., M. Nixon and L. Kutny. 2004. Circumpolar-Active-Layer-Monitoring (CALM) sites in the Mackenzie Valley, northwestern Canada. *Permafrost and Periglacial Processes*, 15:141–153.

Tarnocai, C. 2004. Northern Soil Research in Canada. In John M. Kimble (ed.), *Cryosols: Permafrost-Affected Soils*, Springer-Verlang Publisher, Berlin, Heidelberg, New York, ISBN 3-540-20751-1, pp. 29–43.

Tarnocai, C. 2004. Cryosols of Arctic Canada. In John M. Kimble (ed.), *Cryosols: Permafrost-Affected Soils*, Springer-Verlang Publisher, Berlin, Heidelberg, New York, ISBN 3-540-20751-1, pp. 95–117.

Tarnocai, C. 2004. Classification of Cryosols in Canada. In John M. Kimble (ed.), *Cryosols: Permafrost-Affected Soils*, Springer-Verlang Publisher, Berlin, Heidelberg, New York, ISBN 3-540-20751-1, pp. 599–610.

Tarnocai, C. 2004. Classification of Permafrost-Affected Soils in the WRB. In John M. Kimble (ed.), *Cryosols: Permafrost-Affected Soils*, Springer-Verlang Publisher, Berlin, Heidelberg, New York, ISBN 3-540-20751-1, pp. 637–656.

Tarnocai, Charles. 2004. Development of international databases to manage global land resources. Proceeddings of the Innovative Techniques in Soil Surveys Conference. Edited by: Hari Eswaran, Pisoot Vijarnsorn, Taweesak Vearasilp and Eswaran Padmanabhan. Land development department, Chattuchak, Bangkok, Thailand

Tarnocai, Charles. 2004. Classification, Canadaian. Encyclopedia of Soil Science. Marcel Dekker Publisher, New York, U.S.A. DOI: 10.1081/E–ESS 120025287, pp1-6.

Tarnocai, C. 2004. Organic Carbon in Cryosolic Soils in the Northern Circumpolar Area. CliC Newsletter "Ice and Climate", p. 10-11.

Tarnocai, C., Hohban, L. and. Lacelle, B. 2004. Peatlands of Coastal British Columbia. Agriculture and Agri-Food Canada, Research Branch, Ottawa ON. (digital database).

Tarnocai, C., Hohban, L. and Lacelle, B. 2004. Peatlands of Southern Ontario. Agriculture and Agri-Food Canada, Research Branch, Ottawa ON. (digital database).

Tarnocai, C., Hohban, L. and. Lacelle, B. 2004. Peatlands of Coastal British Columbia. Agriculture and Agri-Food Canada, Research Branch, Ottawa ON. (two 1:600 000 scale maps).

Tarnocai, C., Hohban, L. and Lacelle, B. 2004. Peatlands of Southern Ontario. Agriculture and Agri-Food Canada, Research Branch, Ottawa ON. (four 1:350 000 scale maps).

Tarnocai, C., Kettles, I. M. and Lacelle, B. 2005. Peatlands of Canada Database. Agriculture and Agri-Food Canada, Research Branch, Ottawa, ON. (digital database)

Tarnocai, C., Kettles, I.M. and. Lacelle, B. 2005. Peatlands of Canada. Agriculture and Agri-Food Canada, Research Branch, Ottawa, ON. (1:6 500 000 scale map)

Tarnocai, C., Kettles, I. M. and Lacelle, B. 2005. Soil Organic Carbon Content of Canadian Peatlands. Agriculture and Agri-Food Canada, Research Branch, Ottawa, ON. (1:7 500 000 scale map)

Tarnocai, C., Kettles, I. M. and Lacelle, B. 2005. Soil Organic Carbon Mass of Canadian Peatlands. Agriculture and Agri-Food Canada, Research Branch, Ottawa, ON. (1:7 500 000 scale map)

Steven C. Wofsy
Harvard University, Room 100A, Pierce Hall, 29
Oxford St., Cambridge, MA 02138.
Telephone: 617-495-4566; FAX 617-495-4551

Education

1966	B.S. (with honors) in Chemistry	University of Chicago, Chicago, Illinois
1967	M.A. in Chemistry	Harvard University, Cambridge, MA.
1971	Ph.D. in Chemistry	Harvard University, Cambridge, MA.

Research Interests

Terrestrial carbon cycle; effects of forests on climate, and climate on forests.
Inference of large scale carbon budgets from atmospheric and land surface data
CO_2 as a tracer of atmospheric transport in the upper troposphere and stratosphere
New instrumentation for measuring atmospheric carbon cycle species (CO_2, CO, CH_4).

Employment History

June 1971 to September 1973. NRC Research Associate, Smithsonian Astrophysical Observatory.

September 1973 to June 1977. Division of Engineering and Applied Physics, Harvard, Lecturer and Research Fellow on Atmospheric Chemistry (Harvard DEAS).

July 1977 to June 1982. Associate Professor of Atmospheric Chemistry, (Harvard DEAS).

July, 1982 to February, 1995. Senior Research Fellow, (Harvard DEAS).

February, 1995. Gordon McKay Professor of Atmospheric and Environmental Sciences, Harvard (DEAS) and Department of Earth and Planetary Sciences (EPS).

January, 1997. Abbott Lawrence Rotch Professor of Atmospheric and Environmental Science, Harvard University DEAS and EPS.

Project or Lead Scientist for the following aircraft measurement programs

Stratospheric Photochemistry, Aerosol, and Dynamics Experiment (NASA ER-2, 1992-3)
Stratospheric Tracers of Atmospheric Transport (STRAT; NASA ER-2 platform, 1995-7)
CO_2 Boundary-layer Regional Atmospheric Study (COBRA, UND Citation 2, 1999-2000,
         NASA/NOAA/NSF/DoE)
CO_2 Boundary-layer Regional Atmospheric Study- North American Carbon Program, Canada-US
         Preliminary Study (May - June 2003, NASA/TEP).
CO_2 Boundary-layer Regional Atmospheric Study-Maine (COBRA, U. Wyoming King Air, 2004 NSF/Biocomplexity)

Committees

NASA Earth System Science and Applications Advisory Committee 1995-2000; chair, 1997-1999;   NASA Advisory Council, 1997-1999.

Carbon Cycle Science Plan Working Group, co-chair, 1998-1999; North American Carbon Program writing group, chair, 2001-2003.

Publications

Andrews, A. E., K. A. Boering, S. C. Wofsy, B. C. Daube, D. B. Jones, S. Alex, M. Loewenstein, J. R. Podolske, and S. E. Strahan, Empirical age spectra for the midlatitude lower stratosphere from *in situ* observations of CO_2, *J. Geophys. Res, 106*, 10257-10274, 2001.

Barford, Carol C., Steven C. Wofsy, Michael L. Goulden, J. Wm. Munger, Elizabeth Hammond Pyle, Shawn P. Urbanski, Lucy Hutyra, Scott R. Saleska, David Fitzjarrald, Kathleen Moore, Factors controlling long and short term sequestration of atmospheric CO_2 in a mid-latitude forest, *Science 294 (5547):* 1688-1691, 2001.

Chou, Wendy W., Steven C. Wofsy, Robert C. Harriss, John C. Lin, C. Gerbig, and Glenn W. Sachse, Net fluxes of CO_2 in Amazônia derived from aircraft observations, *i. Geophys Res. 107 (D22), 4614, 10.1029/2001JD001295, 2002.*

Daube BC; Boering KA; Andrews AE; Wofsy SC: A high-precision fast-response airborne CO_2 analyzer for in situ sampling from the surface to the middle stratosphere. *J. Atmos. Oceanic Technol. 19*, Iss 10, pp 1532-1543, 2002.

Goldstein, A. H., S.M. Fan, M.L. Goulden, J.W. Munger, S.C. Wofsy. Biogenic Olefin Emissions from a Midlatitude Forest, *J. Geophys. Res. 101*, . 9149-9157, 1996.

Goulden, M. L., J. W. Munger, S.-M. Fan, B. C. Daube, and S. C. Wofsy, Effects of interannual climate variability on the carbon dioxide exchange of a temperate deciduous forest, *Science 271*, 1576-1578, 1996.

Goulden, M. L., J. W. Munger, S.-M. Fan, B. C. Daube, and S. C. Wofsy, Measurements of carbon storage by long-term eddy covariance, *Global Change Biology 2, 169-182,* 1996.

Gu, Lianhong, Dennis D. Baldocchi, Steven C. Wofsy, J. William Munger, Joseph J. Michalsky, Shawn P. Urbanski, Thomas A. Boden, Response of a deciduous forest to the Mt. Pinatubo eruption: Enhanced photosynthesis, *Science 299*, 2035-2038, 28 March 2003.

Lin, J.C., C. Gerbig, S.C. Wofsy, A.E. Andrews, B.C. Daube, K.J. Davis, A. Grainger, The Stochastic Time-Inverted Lagrangian Transport Model (STILT): Quantitative analysis of surface sources from atmospheric concentration data using particle ensembles in a turbulent atmosphere, *J. Geophys. Res.* 108, No. D16, 4493, *10.1029/2002JD03161,2003.*

Lai, Chun-Ta, James R. Ehleringer, Steve Wofsy, Dave Hollinger, and P.P. Tans. Estimating photosynthetic ^{13}C discrimination in terrestrial CO_2 exchange from canopy to regional scales (accepted in *Global Biogeochemical Cycles*).

Litvak, M., S. Miller, S. Wofsy, M. Goulden, Effect of stand age on whole-ecosystem CO_2 exchange in the Canadian boreal forest. *J. Geophys. Res. Doi:* 10.1029/2001/JD000854, 2003.

Munger, J. William, Song-Miao Fan, Peter S. Bakwin, Mike L. Goulden, A. H. Goldstein, A. S. Colman, and Steven C. Wofsy, Regional budgets for Nitrogen Oxides from Continental Sources: Variations of rates for oxidation and deposition with season and distance from source regions, *J. Geophys.Res., 103*: (D7) 8355-8368, 1998

Potosnak, M. J. S. C. Wofsy, A. S. Denning, T. J. Conway, J.W. Munger, and D. H. Barnes, Influence of biotic exchange and combustion sources on atmospheric CO_2 concentrations in New England from observations at a forest flux tower. *J. Geophys. Res, 104*: 9561-9569, 1999.

     Turner, David P., Shawn P. Urbanski, Dale Bremer, Steven C. Wofsy, Tilden Meyers, Stith T. Gower, Matthew Gregory A Cross-biome Comparison of Daily Light Use Efficiency for Gross Primary Production, *Global Change Biology (in press, 2003)*.

     Wofsy, S. C. and R.C. Harriss, 2002: *The North American Carbon Program (NACP)*. Report of the NACP Committee of the U.S. Interagency Carbon Cycle Science Program. Washington, DC: *US Global Change Research Program*, 75pp.

     Wofsy, Steven C. , Where Has All the Carbon Gone? *Science 292*: 2261-2263. (in Perspectives), 2001.